Giulia Fredi
Thermal Energy Storage Composites

Also of interest

Smart Materials.
Electro-Rheological Fluids, Piezoelectric Smart Materials, Shape Memory Alloys
Kaushik Kumar, Chikesh Ranjan, 2025
ISBN 978-3-11-137901-2, e-ISBN (PDF) 978-3-11-137962-3

Crop Nutrition.
Enhancing Healthy Soils, Food Security, Environmental Sustainability and Advancing SDGs
Amanullah, 2024
ISBN 978-3-11-161709-1, e-ISBN (PDF) 978-3-11-161767-1

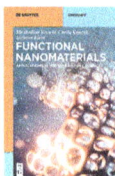

Functional Nanomaterials.
Applications in Medicine and Life Sciences
Stefanie Klein, Maximilian Kryschi, Carola Kryschi, 2024
ISBN 978-3-11-060545-7, e-ISBN (PDF) 978-3-11-060549-5

Giulia Fredi

Thermal Energy Storage Composites

Multifunctional Structural Polymer Composites
for Thermal Energy Storage and Management

DE GRUYTER

Author
Dr. Giulia Fredi
Department of Industrial Engineering
University of Trento
Via Sommarive 9, 38123 Trento, Italy
giulia.fredi@unitn.it

The present work contains ideas and parts of the author's dissertation, which she defended in June 2020 at the Department of Industrial Engineering of the University of Trento. The mentioned thesis is freely accessible at this link: https://iris.unitn.it/retrieve/e3835196-9f15-72ef-e053-3705fe0ad821/Giulia_Fredi_PhD_Thesis.pdf

ISBN 978-3-11-111155-1
e-ISBN (PDF) 978-3-11-111186-5
e-ISBN (EPUB) 978-3-11-111278-7

Library of Congress Control Number: 2024952879

Bibliographic information published by the Deutsche Nationalbibliothek
The Deutsche Nationalbibliothek lists this publication in the Deutsche Nationalbibliografie; detailed bibliographic data are available on the Internet at http://dnb.dnb.de.

© 2025 Walter de Gruyter GmbH, Berlin/Boston, Genthiner Straße 13, 10785 Berlin
Cover image: Olena Lishchyshyna/iStock/Getty Images Plus
Typesetting: Integra Software Services Pvt. Ltd.

www.degruyter.com
Questions about General Product Safety Regulation:
productsafety@degruyterbrill.com

Contents

1 Principles of thermal energy storage (TES)

The efficient storage and management of thermal energy has garnered increasing interest in recent decades, primarily due to its significance in conserving and optimizing energy resources. This chapter starts by delving into the rationale behind energy conservation and management, followed by an exploration of the concept of thermal energy storage and the categorization of various TES technologies [1].

1.1 Importance of energy conservation and management

The increasing worries about the exhaustion of fossil fuels, the effects of climate change, and the emission of greenhouse gases have sparked the interest of researchers, industries, and governments in developing technologies that promote a more efficient use of energy sources. As the strategies for energy production based on fossil fuels have recently been undoubtedly proven to contribute to environmental pollution and global warming, the focus has shifted towards sustainable and renewable energy sources, such as solar, wind, and geothermal energy [2].

The main challenges preventing the widespread adoption of renewable energy sources are their intermittent nature and the high initial investments needed for their exploitation. Although technological advancements may help reduce initial plant costs, addressing the intermittency of these sources requires enhancing energy storage technologies. Renewable energy sources are subject to fluctuations in output due to daily and seasonal variations and/or unpredictability, which necessitates the integration of energy storage systems to maintain a consistent output during off-peak periods [2].

Energy storage systems can provide advantages in the utilization of conventional energy sources when applied in buildings, vehicles, and industrial applications [2, 3]. For instance, in the transportation industry, improved battery performance can encourage the adoption of electric vehicles, thereby decreasing the demand for traditional fuels. Energy storage systems can help decrease equipment sizes and initial and maintenance costs, boost plant flexibility and efficiency, and reduce the need for emergency power generators that consume primary energy sources. All of these factors can lower overall energy consumption and costs [1, 3].

1.2 Energy storage technologies

Energy storage systems can be classified according to the form of intermediate energy, which can be [2]:
- chemical (e.g., hydrogen storage);
- electrical (e.g., capacitors);

https://doi.org/10.1515/9783111111865-001

- electrochemical (e.g., batteries, fuel cells);
- mechanical (e.g., compressed air);
- thermal (e.g., sensible heat storage in water tanks).

Other classifications consider different aspects of energy storage, such as the type of input energy (e.g., electrical, thermal) or output energy (e.g., thermal energy, liquid fuels), as well as the specific conversion process used (e.g., power-to-power, power-to-gas, power-to-liquid). Although these different forms of energy storage have varying power rates, discharge rates, efficiencies, and levels of technological maturity, they all offer significant benefits in their respective applications. By maintaining a continuous power supply and enhancing system performance and reliability, they can have a considerable positive impact [2]. The following sections focus on thermal energy storage and explore the technologies and applications associated with this concept.

1.3 Concept of thermal energy storage

Thermal energy storage (TES) refers to the temporary storage of heat, which can be utilized at a later time or in a different location. The primary objective of TES systems is to bridge the gap between the availability and demand of thermal energy, thereby facilitating the recovery of waste heat and enhancing the utilization of intermittent energy sources [4]. Unlike other energy storage systems, the storage of energy in the form of heat (or cold) has longer storage times and higher efficiency [2]. The operation of a typical TES system, depicted in Fig. 1.1, involves charge storage, and discharge steps.

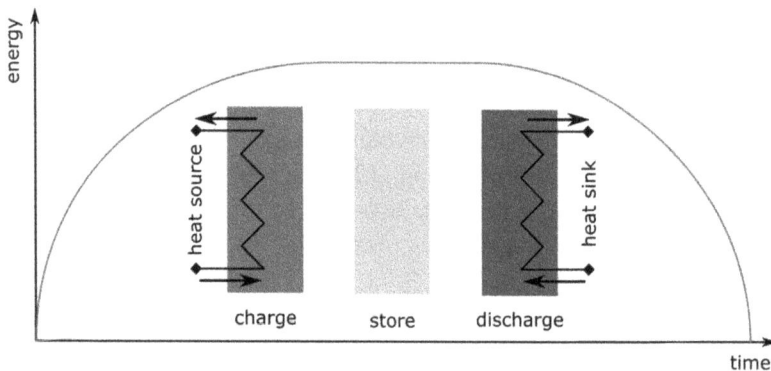

Fig. 1.1: Typical working cycle of a TES system (adapted from [2] and [5]).

TES materials and technologies have gained considerable attention, as evidenced by the substantial increase in the number of scientific publications on the subject.

Figure 1.2 illustrates the yearly count of publications retrieved using the search phrase "thermal energy storage" on the Web of Science database. Between 2008 and 2023, the number of scientific studies conducted has increased over 20 times, highlighting the significant efforts made by the scientific community to expand knowledge and technologies in this field.

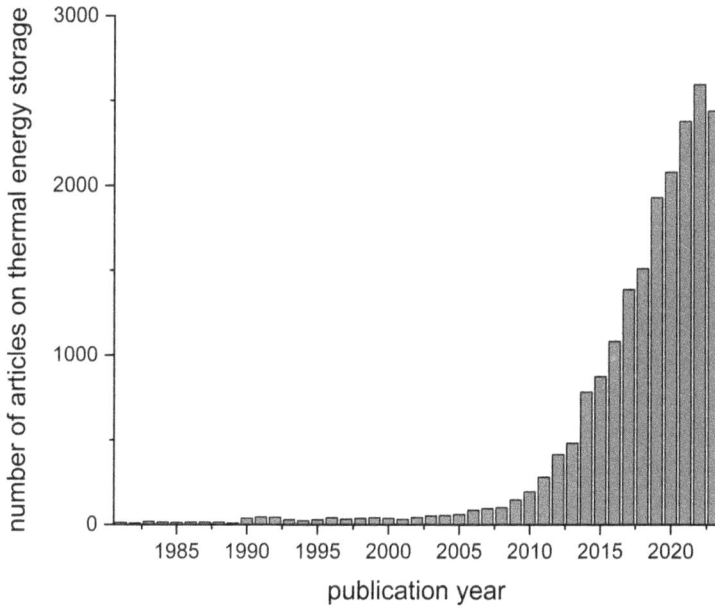

Fig. 1.2: Number of scientific papers published yearly about thermal energy storage from 1980 to 2023 (ISI Web of Science database, consulted on April 12, 2024).

TES technologies are currently employed for three main purposes:
(1) to store waste/excess heat that will be released during off-peak periods, e.g., to recover waste industrial heat [6], or in solar thermal power plants [7];
(2) to contribute to temperature regulation, e.g., in buildings, to store excess energy during the day and release it during the night or in off-peak times [8], or for body temperature regulation through smart thermoregulating garments [9];
(3) to temporarily store heat and prevent a temperature rise that would otherwise damage a component, as in the thermal management of electronic devices [10].

Based on this classification, it is evident that, in certain situations, the desired product is the stored and released thermal energy, as seen in cases (1) and (2). In these instances, the excess energy is saved for later use, which can involve either temperature regulation, as in the case of buildings, or an increase in efficiency, as in the case of power plants. These instances are generally referred to as "thermal energy storage or

TES *properly said*," and they typically require energy storage systems with high thermal capacity to store as much energy as possible. In contrast, in other situations, the excess heat is not stored for later use but is instead employed to prevent a dangerous rise in temperature, as in case (3). These instances are examples of "thermal management" (TM), and their primary requirement is usually a well-defined, and generally high, energy storage rate. However, it can be challenging to distinguish between TES properly said and TM; some other classifications categorize only case (1) as TES properly said, while the other cases are examples of TM, as the main goal is to keep the temperature within an optimal range. This book will explore both TES and TM without making a strict distinction, although the multifunctional composites described in the following chapters would be more suitable for thermal management applications.

1.4 Classification of TES technologies

TES technologies are categorized based on the method utilized to vary the internal energy of the storage medium (Fig. 1.3) [11]. Thermal energy can be stored and released by changing the temperature of a material (*sensible heat TES*, SH-TES), through an endo/exothermic phase change (*latent heat TES*, LH-TES), or through a thermochemical reaction (*thermochemical heat TES*, TH-TES). The selection of a TES system depends on various factors, such as the required heat storage period (hours, months, days), economic considerations, working temperature, and available volume. A general overview of the main parameters for each TES system is provided in Tab. 1.1 [12]. Although TH-TES systems offer a higher energy density per unit mass and volume and improved efficiency, their technological maturity is currently lower than that of the other two classes. Recent reviews on the topic suggest that significant effort should be devoted to studying advanced materials and developing efficient prototypes [13]. In contrast, SH-TES and LH-TES systems such as underground thermal energy storage and domestic hot water storage have low risk and high commercialization potential [2].

The following sections summarize the characteristics and the governing equations of each of the three TES classes, while Chapter 2 focuses on the description of latent heat TES materials and technologies, as they are the focus of this book.

1.4.1 Sensible heat storage (SH-TES)

SH-TES is performed by increasing or decreasing the temperature of the storage medium, and the enthalpy variation is proportional to the temperature difference, as depicted in Fig. 1.4a. More specifically, the total amount of energy ΔE (J) stored in the system can be defined by eq. (1.1), as

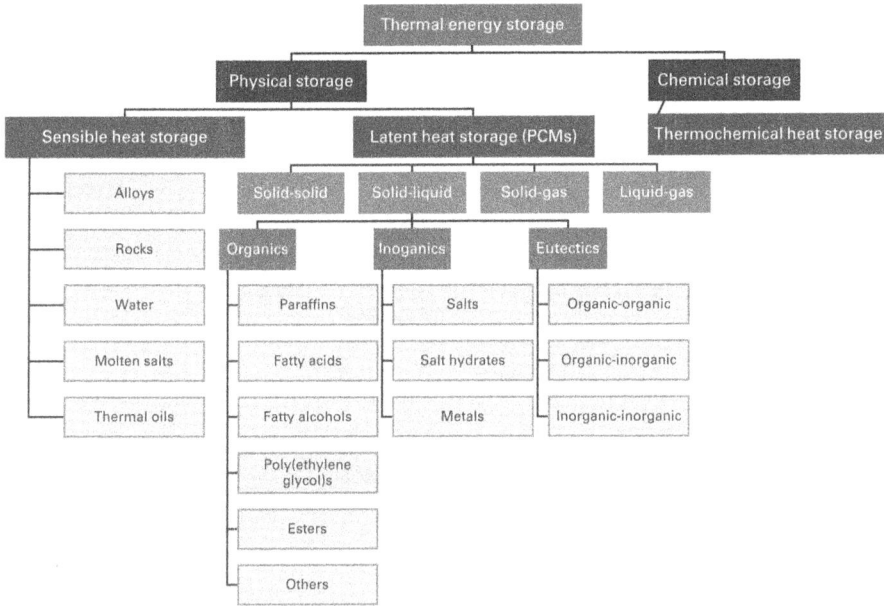

Fig. 1.3: Classification of TES technologies.

Tab. 1.1: General overview of the main performance parameters of sensible, latent, and thermochemical heat TES technologies (adapted from [12]).

TES system	Capacity (kWh/t)	Efficiency (%)	Storage period (h, d, m)	Cost (€/kWh)
Sensible heat storage	10–50	50–90	d/m	0.1–10
Latent heat storage	50–150	75–90	h/m	10–50
Thermochemical heat storage	120–250	75–100	h/d	8–100

$$\Delta E = \int_{H_1}^{H_2} mdH = m(H_2 - H_1) = \int_{T_1}^{T_2} mcdT = mc(T_2 - T_1), \tag{1.1}$$

where m is the mass of the storage medium, H_1 and H_2 the initial and final enthalpy values [J/g], c the specific heat capacity [J/(g·K)], and T_1 and T_2 the initial and final temperature [°C]. An analogous equation can be written for the energy release, which results in the cooling of the storage medium. From eq. (1.1), it is clear that the efficiency of a storage medium is influenced by the available mass and volume, and the energy stored per unit mass increases with the specific heat capacity. Additionally, it is important that the medium is non-toxic, cost-effective, and maintains stability over numerous thermal cycles. Furthermore, the medium should be well insulated, as any changes in the surrounding temperature affect the stored energy, and it should have a high thermal diffusivity, which increases the heat transfer rate within the medium and the overall thermal exchange rate.

Typical materials used as sensible heat storage media are liquids such as water, oils, or molten salts, or solids such as metals or rocks [14]. One of the most widespread sensible heat storage media is water, due to the high specific heat, high availability, and low cost. It can be used over a wide temperature range (0–100 °C) and serves both as a storage and transport medium. It is the most commonly used storage medium for solar-based hot water systems and radiation systems for indoor heating [14].

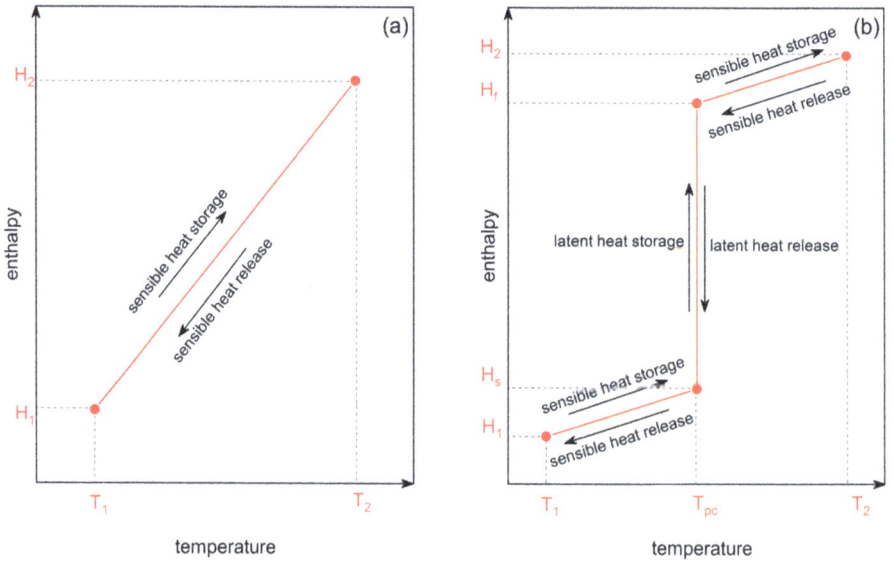

Fig. 1.4: Variation of enthalpy of the storage medium as a function of its temperature in (a) an SH-TES unit cycle and (b) in an LH-TES unit cycle (adapted from [2] and [5]).

1.4.2 Latent heat storage (LH-TES)

LH-TES employs phase transitions for heat storage and release. Most LH-TES systems utilize the melting-solidification phase transition, where the storage medium absorbs heat during melting and releases it upon solidification. Evaporation-condensation phase changes are typically avoided due to the significant volume variation, which increases the requirements and complexity of the confinement units.

The enthalpy variation as a function of temperature for a solid to-liquid phase transition is illustrated in Fig. 1.4b. During an energy storage process, the medium initially behaves like an SH-TES unit and absorbs sensible heat, with a consequent increase in temperature. This behavior continues until the phase-change temperature

(T_{pc}) is reached. Then, the material undergoes a phase change and remains at T_{pc} until the completion of the phase transition, during which the absorbed energy is equal to the latent heat of phase change ($\Delta H_{pc} = H_f - H_s$). Further increases in stored enthalpy rise the temperature of the storage medium; the slope of the enthalpy-temperature relationship depends on the specific heat capacity of the substance after phase change and can be different from that between T_1 and T_{pc}. The total enthalpy variation ΔE [J] of such process is described by eq. (1.2), as

$$\Delta E = \int_{H_1}^{H_2} mdH = m(H_2 - H_1) = \int_{T_1}^{T_{pc}} mc_s dT + m\Delta H_{pc} + \int_{T_{pc}}^{T_2} mc_L dT, \tag{1.2}$$

where ΔH_{pc} [J/g] is the latent heat of phase change, T_{pc} is the phase change temperature, and c_s and c_L are the specific heat capacity of the solid and liquid phases, respectively. Generally, the latent heat is considerably higher than the sensible heat absorbed, which means that LH-TES systems normally require less material use and less volume availability, as they can store a high amount of energy in a smaller weight and volume.

1.4.3 Thermochemical energy storage (TH-TES)

The third type of TES system consists of storing and releasing heat in a reversible endo/exothermic thermochemical reaction. During the endothermic stage (charging), the storage medium absorbs heat from the environment as the reaction enthalpy and it typically splits into two or more chemical substances, as illustrated in eq. (1.3), as

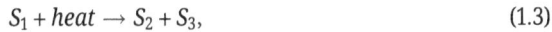

$$S_1 + heat \rightarrow S_2 + S_3, \tag{1.3}$$

where S_1 represents the storage medium and S_2 and S_3 the reaction products, which can be stored separately for a long time. The reverse reaction constitutes the discharging process, where the reaction between S_2 and S_3 releases the same amount of heat stored during the endothermic stage. The total heat released depends on the reaction enthalpy [J/mol] and the quantity of material [mol].

The potential of thermochemical heat storage is noteworthy, as TH-TES materials have up to 10 times higher energy storage density than SH-TES media, and approximately two times higher than the most common LH-TES materials (Fig. 1.5) [15, 16]. This aspect and the remarkably low heat losses are the two main advantages of TH-TES systems. However, the full exploitation of the potential of these materials is possible only with efficient heat and mass transfer to and from the storage volume, which can put limitations on the maximum size of the storage volume itself. Achieving such efficient heat and mass transfer and coping with slow reaction kinetics are the key targets of the research on TH-TES systems, as they are the main issues to be overcome for the scale-up and commercialization of these systems [15, 17].

TH-TES materials can be further classified as chemical or sorption systems. Chemical systems are those in which a considerable amount of heat is generated from an exothermic synthesis reaction and their working principle is properly described by the model reaction reported in eq. (1.3). In the sorption systems, a gas, the sorbate, reacts with a sorbent, which can be solid (absorption reactions) or liquid (adsorption reactions) [15]. Sorption systems are characterized by faster kinetics and lower working temperatures than chemical systems, but the energy storage density is also lower [17].

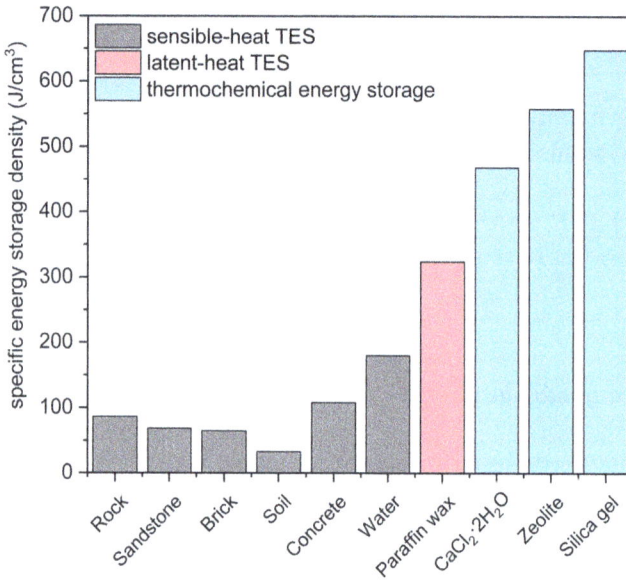

Fig. 1.5: Comparison of specific energy storage density values of different sensible, latent and thermochemical TES media (adapted with permission from [15]).

2 Latent heat TES and phase change materials (PCMs)

The current chapter delves into the intricacies of latent heat TES (LH-TES). It explicates the working principles and benefits of latent heat storage systems in greater detail and elucidates the categories of LH-TES materials, confinement methods, and the cost and environmental implications that may arise from their utilization. As the melting-crystallization phase change is indisputably the most commonly employed transition in LH-TES technology, the ensuing discussion will also refer to this specific transition using the more general terms "phase change" or "phase transition," unless explicitly stated otherwise.

2.1 Principle and advantages of latent heat TES

Latent heat TES has gained popularity over other TES technologies for a multitude of applications, primarily due to three key advantages.

First, LH-TES can store and release a substantial amount of heat per unit mass or volume, which is attributable to the high energy density of phase change materials (PCMs). This high energy density allows LH-TES systems to be much more compact compared to sensible heat storage using water tanks. It enables the use of significantly less material compared to traditional sensible heat storage systems, thereby enhancing system flexibility and design possibilities while reducing initial and maintenance costs.

The second benefit is related to the constant temperature storage and release of heat, specifically the phase change temperature (T_{pc}). During the energy storage phase (charging), the PCM reaches the melting point and maintains a constant temperature throughout the entire melting process, regardless of the heat flux applied or slight fluctuations in the surrounding temperature. This same scenario occurs during heat release (discharging): when the PCM approaches the crystallization phase change, the temperature remains constant until the transition is completed. This feature is appealing in thermal management applications where temperature stability is crucial, such as maintaining a specific temperature range (e.g., 20–25 °C for indoor temperature control in buildings) or keeping the temperature below a critical level (e.g., below 50 °C for the thermal management of electronic devices).

The third advantage is represented by the technological maturity of LH-TES systems, which often makes LH-TES a preferred choice over thermochemical heat storage techniques. Even though TH-TES systems exceed LH-TES in energy storage density, solid-liquid phase change materials are generally easier to handle and exhibit a little volume variation as they do not involve gas phase, thereby requiring smaller systems and less support equipment like compressors or vacuum chambers.

https://doi.org/10.1515/9783111111865-002

2.2 Selection and properties of a phase change material

Selecting the appropriate phase change material is crucial for optimizing the performance and efficiency of a latent heat thermal energy storage (LH-TES) system [18]. With a vast array of PCM options available and continuous research into developing new materials, it is essential to identify the key properties that align with the specific application requirements (Tab. 2.1).

One of the most significant properties, and the primary criterion for selecting a PCM, is the phase change temperature. This temperature should be below that of the heat source but above that of the working environment, because a PCM already melted before absorbing heat from the designated source would be purposeless. For example, in solar energy applications, the PCM melting point should be around 200–400 °C to match the operating temperatures of concentrated solar power plants. Conversely, for building heating and cooling, PCMs with melting points between 20 °C and 30 °C are preferred to store energy from solar thermal collectors or off-peak electricity. For applications aimed at preventing overheating, such as electronics cooling, the ideal PCM melts slightly below the maximum allowable temperature. This strategic melting point slows the rate of phase change, extending the thermal management window.

The phase change enthalpy, quantifying the energy stored/released per unit mass or volume during phase transition, is another pivotal property. High phase change enthalpy values result in smaller system sizes, as less material is required to store a certain amount of energy. Organic PCMs like paraffin waxes exhibit high enthalpies around 200–250 kJ/kg, enabling compact TES designs. Inorganic salt hydrates like $Na_2SO_4 \cdot 10H_2O$ have even higher enthalpies exceeding 300 kJ/kg. The PCMs with high melting enthalpy generally also exhibit high specific heat capacity, which is desirable as it increases the sensible heat stored before and after the melting temperature range, thereby enhancing the overall energy exchange.

Efficient heat transfer within the PCM is vital, making thermal conductivity a key factor for both solid and liquid phases. While most organic PCMs have low intrinsic conductivities around 0.1–0.3 W/(m · K), incorporating thermally conductive additives like graphite, metallic fins/foams, or encapsulating the PCM can significantly improve heat distribution throughout the storage material and away from the heat source, thus avoiding heat bottlenecking. The density and density variation between the melted and the solid state are also factors to be considered in the processes of material selection and system design. Additionally, for long-term cycling, PCMs must demonstrate physical/chemical stability, compatibility with containment materials, non-corrosiveness, and non-toxicity. Salt hydrates are generally stable but require careful encapsulation to prevent deleterious reactions. A good PCM should also exhibit congruent melting, low supercooling degree, and completely reversible melting/crystallization cycles.

While no currently available PCM possesses the complete set of optimal properties, researchers continue enhancing existing materials and exploring new composites/nanostructured PCMs. Phase change temperatures can be tuned by forming eutectics, while micro/nano-encapsulation techniques prevent leakage and may also improve conductivity. These advances open up possibilities for tailoring PCM properties to meet demanding application requirements.

Tab. 2.1: Summary of the most important properties of PCMs used as selection criteria.

Property	Desirable values/examples
Phase change temperature	In the right temperature range for the application: e.g., 20–30 °C for building heating/cooling, 200–400 °C for concentrated solar power applications, just below max allowable temperatures for thermal management
Phase change enthalpy	High values desirable, e.g., 200–250 kJ/kg for paraffin waxes, >300 kJ/kg for salt hydrates like $Na_2SO_4 \cdot 10H_2O$
Specific heat capacity	High values enhance sensible heat storage before/after phase change.
Thermal conductivity	High values desirable. Most organic PCMs have low intrinsic conductivity (0.1–0.3 W/ m · K), which requires enhancement via additives, fins, encapsulation, etc.
Density and density variation	High density and small density variation changes across phase change desirable
Cycle stability	High cycle stability required for long-term applications
Chemical stability	High chemical stability desirable
Compatibility	Non-corrosive, compatible with metals, polymers, etc.
Toxicity	Non-toxic and non-hazardous preferred
Melting behavior	Congruent melting, minimal/no supercooling, fully reversible phase changes desired
Availability	Large availability desired
Cost	Low cost desired

2.3 Classification of PCMs

The number of materials currently used as PCMs is noteworthy and constantly growing, which makes it challenging to develop a fully comprehensive list. This section aims to describe a commonly accepted way to classify the PCMs, illustrating their working principles, advantages, and disadvantages. As already described in Chapter 1 (Fig. 1.3), solid-liquid PCMs can be divided into organic PCMs, inorganic PCMs, and eutectic mixtures of organic and/or inorganic PCMs [19].

The following sections describe the characteristics and advantages of each PCM class.

2.3.1 Organic PCMs

Organic PCMs are the most widely used PCMs in the low-medium temperature range (0–100 °C) and are diffused in the thermal management of buildings and electronic devices. They are an affordable and convenient option, with a relatively high energy density, and come in a variety of molecular weights, allowing for a wide range of working temperatures. They are non-toxic, do not release volatile substances, and are characterized by congruent melting and negligible supercooling. However, their thermal conductivity is generally low, which can be addressed by increasing the heat transfer area, using highly thermally conductive containers, or adding metallic or carbon-based micro/nano-fillers [19]. Additionally, organic PCMs are flammable due to their hydrocarbon nature, but their flash point is approximately 200 °C, which is well above the operating temperature range. Organic PCMs comprise paraffin waxes, poly (ethylene glycol)s, and fatty acids, but also other compounds such as ketones, esters, ethers, halogen derivatives, sulfur compounds, and oleochemical carbonates. Some of the most common organic PCMs are listed in table Tab. 2.2.

Paraffin waxes. The most widely used organic PCMs are paraffin waxes, saturated hydrocarbons with the chemical formula C_nH_{2n+2}. Paraffins embody all the aforementioned advantages of organic PCMs, and they are generally cheaper and exhibit higher heat of fusion (200–240 J/g) and specific heat capacity (2.1–2.4 kJ/(kg · K)) than other organic compounds [20]. They present superior thermal stability over repeated thermal cycles (also after 1,000–2,000), are largely commercially available, and have a small volume change during phase change and a relatively low vapor pressure [21]. Paraffins are available with a broad range of chain lengths; those between C5 (pentane) and C15 (pentadecane) are liquid at room temperature, while those containing a higher number of carbon atoms are solid with a waxy appearance. Commercial paraffins are a blend of various hydrocarbons that do not exhibit phase separation even after many thermal cycles, and their formulation is designed to select the desired melting temperature and the highest possible melting enthalpy. These paraffins are typically resistant to both chemical and environmental degradation, although they can undergo slow oxidation when exposed to oxygen, which has led to the use of sealed containers to prevent this reaction [22].

Fatty acids. Fatty acids, represented by the chemical formula $CH_3(CH_2)_{2n}COOH$, are the second most popular organic solid-liquid PCM. They are divided into six classes: lauric, myristyl, caprylic, capric, palmitic, and stearic. Fatty acids show high melting enthalpy values (45–210 J/g) and a wide range of melting temperatures (−5 °C to 70 °C) but are three times more expensive than paraffins. As they can be produced from vegetal or animal bio-sources and are biodegradable, they have been the subject of extensive studies to replace paraffins for low/medium temperature applications, such as solar energy storage and thermal management of indoor environments [20].

Poly(ethylene glycol)s (PEGs). PEGs, also known as poly(ethylene oxide)s (PEOs), are composed of dimethyl ether chains HO-CH$_2$-(CH$_2$-O-CH$_2$-)$_n$-CH$_2$-OH. Due to the amphiphilic nature of their chain, which presents hydrocarbon sequences and polar groups such as -OH, PEGs are soluble in water and in some organic solvents [23]. Also for PEGs, the melting temperature and enthalpy increase with the molecular weight; for example, PEG600 melts at 18.5 °C absorbing 121.1 J/g, while PEG2000 has a melting temperature of 61.2 °C and a melting enthalpy of 176.2 J/g. As for the other organic PCMs, the low thermal conductivity is an issue that must be considered in the applications [24]. PEGs are biodegradable and biocompatible and are also used in drug delivery systems. This feature has expanded the use of PEG as a PCM in applications inside the human body, e.g., to subtract heat during the in-situ polymerization of acrylic bone cement and avoid overheating and damage to the surrounding biological tissues [25].

Tab. 2.2: List of commonly used organic phase change materials with their melting temperature and enthalpy.

Category	Substance(s) and ratio	Melting temperature (°C)	Melting enthalpy (kJ/kg)	Ref.
Paraffins	n-Octadecane (C$_{18}$H$_{38}$)	28	244	[26]
	n-Eicosane (C$_{20}$H$_{42}$)	37	247	[19]
	n-Docosane (C$_{22}$H$_{46}$)	44	249	[19]
	n-Tetracosane (C$_{24}$H$_{50}$)	50	255	[19]
	n-Hexacosane (C$_{26}$H$_{54}$)	56	256	[19]
	n-Octacosane (C$_{28}$H$_{58}$)	61	202	[23]
	n-Triacontane (C$_{30}$H$_{62}$)	65	252	[19]
	n-Dotriacontane (C$_{32}$H$_{66}$)	70	170	[18]
Fatty acids	Caprylic acid CH$_3$(CH$_2$)$_6$COOH	17	148	[18]
	Capric acid (CA) CH$_3$(CH$_2$)$_8$COOH	32	153	[26]
	Lauric acid (LA) CH$_3$(CH$_2$)$_{10}$COOH	44	178	[26]
	Myristic acid (MA) CH$_3$(CH$_2$)$_{12}$COOH	54	195	[19]
	Palmitic acid (PA) CH$_3$(CH$_2$)$_{14}$COOH	64	185	[26]
	Stearic acid (SA) CH$_3$(CH$_2$)$_{16}$COOH	69	202	[26]
Monohydroxy alcohols	1-Dodecanol (C$_{12}$H$_{26}$O)	18	186	[19]
	1-Tetradecanol (C$_{14}$H$_{30}$O)	39	221	[19]
Esters	Methyl palmitate (C$_{17}$H$_{34}$O$_2$)	27	163	[19]
	Butyl stearate (C$_{22}$H$_{44}$O$_2$)	19	140	[18]
Polyethylene glycols	PEG 600	19	121	[18]
	PEG 1,000	40	168.6	[18]
	PEG 10,000	66	171.6	[18]
	PEG 100,000	67	175.8	[18]

2.3.2 Inorganic PCMs

Inorganic PCMs generally show higher density than organic PCMs, and therefore, even though they exhibit similar enthalpy per unit mass, they can have a remarkably higher enthalpy per volume, thereby allowing the production of more compact TES systems. Moreover, the thermal conductivity of inorganic PCMs can be several times higher than that of their organic counterpart. For these reasons, they are the preferred choice in the medium/high-temperature range (100–1,000 °C) and when there are no strict requirements on non-corrosiveness. Inorganic PCMs comprise several classes of materials, such as salts, salt hydrates, and metal alloys [26] (Tab. 2.3).

Salts and salt hydrates. Salts and salt hydrates have similar molecular structures, but in the case of salt hydrates the crystalline lattice is not so closely packed and can easily host water molecules. Common salts and salt hydrates used as PCMs are $NaNO_3$, KNO_3, KOH, $MgCl_2$, $NaCl$, $MgCl_2 \cdot 6H_2O$, $CaCl_2 \cdot 6H_2O$, and $Na_2SO_4 \cdot 10H_2O$, also called Glauber's Salt [26].

   While salts undergo a proper melting/crystallization behavior at the transition temperature, for salt hydrates the solid-liquid phase change is a dehydration/hydration process of the compound, which decomposes into an anhydrous salt (or a lower hydrate) and water molecules. Their higher vapor pressure, which increases with the hydration degree, can cause a loss of water and a change in the thermal behavior of the compound. Salt hydrates often suffer from supercooling problems: above the dehydration temperature, the anhydrous salt may experience segregation and settle at the bottom of the container due to its higher density, thereby hindering the rehydration process. This issue is usually overcome by stirring, by adding excess water to favor solubilization of the whole mass of anhydrous salt and prevent precipitation, or by adding a thickening agent (e.g., borax, graphite) that reduces the extent of phase separation and often acts as a nucleating agent [27, 28].

   The phase change temperature of these compounds ranges from 10 °C to 900 °C. However, for applications where a melting point up to 70–80 °C is required, organic PCMs are often preferred due to their lower cost, easier handling, lower vapor pressure, superior long-term stability, lower supercooling, and lower tendency to incongruent melting [28].

Metal alloys. Metals have not been extensively investigated as PCMs so far, but they are starting to attract considerable attention thanks to their high thermal conductivity and their stability at high temperatures. The most promising metallic PCMs are cesium, gallium, indium, tin, and bismuth for low-temperature applications, and zinc, magnesium, and aluminum for applications at higher temperatures. Metallic PCMs cover a broad range of melting temperatures, from 28 °C for neat cesium to 661 °C for aluminum, but they are not widely used for low temperature applications due to their low phase change enthalpy. Despite their high density, which partially offsets the limited absolute enthalpy and determines a high enthalpy per unit volume, their physical and TES properties do not match those of the most common organic PCMs. On the other hand, high-melting metals such as Al and Mg alloys also exhibit a considerable phase change enthalpy (350–500 J/g), which makes them attractive for high-temperature solar heating applications, in replacement of inorganic salts that are thermally unstable and prone to phase segregation [26].

Tab. 2.3: List of commonly used inorganic phase change materials with their melting temperature and enthalpy.

Category	Substance(s) and ratio	Melting temperature (°C)	Melting enthalpy (kJ/kg)	Ref.
Metals	Cesium	28	16	[26]
	Gallium	30	80	[26]
	Indium	157	29	[26]
	Tin	232	60	[26]
	Bismuth	271	53	[26]
	Zinc	419	112	[26]
	Magnesium	648	365	[26]
	Aluminum	661	388	[26]
Metal alloys	Al59-35 Mg-6Zn	443	310	[26]
	Al54-22Cu-18 Mg-6Zn	520	305	[26]
	Al65-30Cu-5Si	571	422	[26]
	Al88-Si12	576	560	[26]
Salts	Sodium nitrate $NaNO_3$	307	172	[26]
	Potassium nitrate KNO_3	333	266	[26]
	Magnesium chloride $MgCl_2$	714	452	[26]
	Sodium chloride) NaCl	802	492	[26]
	Potassium fluoride KF	857	452	[26]
Salt hydrates	Magnesium chloride hexahydrate $MgCl_2 \cdot 6H_2O$	117	169	[26]
	Calcium chloride hexahydrate $CaCl_2 \cdot 6H_2O$	29	170–192	[26]
	Glauber's salt $NaSO_4 \cdot 10H_2O$	32	251	[26]

2.3.3 Eutectic PCMs

Eutectic PCMs, some of which are listed in Tab. 2.4, are mixtures of organic and/or inorganic compounds that melt and solidify congruently. They present a sharp melting point and a high phase change enthalpy, and their properties can be finely tailored to meet the requirements of a specific application. They are completely miscible in the molten state and freeze forming an intimate mixture of crystals [27], which ac-

Tab. 2.4: List of commonly used eutectic phase change materials with their melting temperature and enthalpy.

Category	Substance(s) and ratio	Melting temperature (°C)	Melting enthalpy (kJ/kg)	Ref.
Organic/organic	CA-LA 64:36	28	244	[29]
	CA-MA 78.39:21.61	37	247	[29]
	CA-PA 89:11	28	145	[29]
	CA-SA94.47:5.53	30	156	[29]
	MA-PA 64.96:35.04	45	152	[29]
	PA-SA 62.99:37.01	54	179	[29]
	MA-SA 64:36	44	182	[29]
	CA-PA 76.5:23.5	23	156	[29]
	CA-MA-SA 72.5:22.5:5.0	24	159	[29]
	CA-PA-SA 79:13:8	20	129	[29]
	SA-PA-LA 6.77:20.97:72.26	32	159	[29]
Inorganic/inorganic	66.6% $CaCl_2$ $6H_2O$ + 33:3% $MgCl_2$ $6H_2O$	25	127	[30]
	58.7% $Mg(NO_3)$ $6H_2O$ + 41:3% $MgCl_2$ $6H_2O$	59	132	[30]
	66.6% urea + 33:4% NH_4Br	76	161	[30]

counts for a phase transition without segregation. As they are generally designed for a target application, they are usually more expensive than the other classes of PCMs.

The melting temperatures and enthalpies can vary slightly depending on the purity and exact composition of the organic PCM. Additionally, the crystallization temperature may differ from the melting point due to the degree of supercooling required to initiate crystallization. A more comprehensive list of organic, inorganic, and eutectic PCMs is available elsewhere in the literature [19, 30, 31].

Additionally, Kulish and coworkers [18] have recently presented a new library (database) containing information on around 500 different PCMs. The library includes various types of organic and inorganic PCMs with a wide range of operating temperatures. For each PCM, up to nine properties are listed, including phase change temperature, solidification temperature, maximum operating temperature, density, latent heat capacity, specific heat capacity, thermal conductivity, cycleability, and ignition temperature. The authors also propose a new PCM selection method based on calculating the Rényi entropy for a given set of desired criteria. This method requires no subjective judgments and automatically preselects a few top candidate PCMs based solely on numerical values of their physical properties. The main advantages of this entropy-based approach are that it avoids biases from subjective weightings of different criteria and allows all criteria to contribute more uniformly to the selection process. The paper provides a representative sample of the PCM library data and illustrates the application of the Rényi entropy selection method through an example case study. The developed PCM library and selection tool are intended to assist engineers and researchers in identifying suitable PCMs for their specific thermal energy storage applications.

2.4 Confinement techniques for organic PCMs

One of the major drawbacks of PCMs is the need to be confined to avoid leakage and loss of material above the melting temperature. The confinement techniques can be divided into two main groups: (i) encapsulation methods and (ii) shape-stabilization methods. This section will focus on the confinement techniques for organic PCMs, with particular attention to the micro/nano-encapsulation techniques and the shape-stabilization with nanofillers, because they are the most suitable and widely used techniques to embed a PCM in a polymer matrix.

2.4.1 Encapsulation

Encapsulation methods involve a container that physically separates the PCM from the surrounding environment, is stable throughout the entire working temperature range, and accommodates the phase transition and the associated volume change. The containers can be of various sizes, shapes, and materials; one can talk about macro-, micro-, or nano-encapsulation.

Macro-encapsulation is the simplest method of confining a PCM, as it involves the use of a box or a tank made of a thermally conductive material (e.g., aluminum, stainless steel) that is chemically compatible with the PCM. The container should be properly sealed to avoid the leakage of even the least viscous PCMs, and the design should always consider the volume expansion and contraction during the phase change. When there are no strict requirements on the strength, thin flexible plastic (e.g., polyethylene) bags can also be used, as they accommodate the volume change and do not require ullage space.

Micro- and nano-encapsulation are interesting as they allow avoiding bulky containers and feature microbeads with a polymeric or inorganic shell and a PCM core. This confinement technique, with capsules in the micron- or sub-micron-scale range, offers two main advantages. The first is that a microencapsulated PCM is easy to handle and embed in other materials such as gypsum and concrete by simple mixing, and it can also be added to liquids to produce PCM-enhanced heat transfer fluids. The second advantage is represented by the augmented specific surface area (SSA), which increases the heat transfer surface and enhances the overall thermal exchange. The capsule shells must be stable over many melting/solidification cycles and must not have any chemical interaction with the PCM.

There are several physical, physical-chemical, and chemical techniques available to produce PCM microcapsules (Tab. 2.5). Among all techniques, the most diffused, researched, and developed on an industrial scale is in-situ polymerization, which includes interfacial, suspension, and emulsion polymerization [32–35]. These techniques differ from each other mainly in terms of polarity and solubility of the monomers and the initiator. A schematic representation of these techniques is depicted in Fig. 2.1. In the interfacial polymerization, the polymeric shell wall is the result of the polymerization of polar and non-polar monomers dissolved in the water (solvent) and oil (PCM) phase of an oil-in-water emulsion, respectively. The shell growing at the interface of the two phases becomes a barrier to diffusion and limits the reaction kinetics, thereby influencing the shell thickness and morphology. Common shell materials are polyurea, urea-formaldehyde, and melamine-formaldehyde. In suspension polymerization, all the reactants are liposoluble and are dispersed in the water-based medium due to continuous agitation and the help of surfactants. Therefore, with this technique, it is difficult to encapsulate hydrophilic PCMs such as PEGs or salt hydrates. The shape and size of the resulting particles are strongly influenced by the stirring speed, amount of stabilizer, fraction of the monomer phases, and relative viscosity of the droplets and the water medium. This technique is similar to emulsion polymerization; the main difference is the hydrophilic nature of the initiator, which is dissolved in the water phase. Common shell materials for both these techniques are acrylic and styrenic polymers such as poly(methyl methacrylate) (PMMA), polystyrene, and styrene-divinylbenzene copolymers [36, 37].

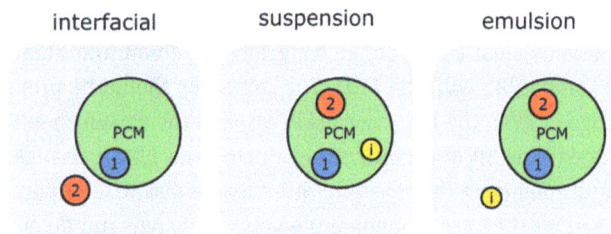

Fig. 2.1: Schematic representation of the chemical microencapsulation methods, i.e., interfacial polymerization, suspension polymerization, and emulsion polymerization (1 and 2: monomers; i: initiator).

Another interesting technique to prepare microcapsules is the sol-gel method, in which a solid shell forms through the gelation of a colloidal suspension (the "sol"). This colloidal suspension, which is often more accurately a solution, is prepared starting from a molecular precursor, such as a metal alkoxide [$M^{n+}(OR)_n$] [38, 39]. Figure 2.2 illustrates the general sol-gel encapsulation route to obtain a silica (SiO_2) shell starting from tetraethyl orthosilicate (TEOS) as the molecular precursor. The PCM is first dispersed in an aqueous medium with the help of surfactants to form a stable oil-in-water (O/W) emulsion. The amount of PCM, the polarity of the aqueous medium, and the type and concentration of surfactant are important parameters determining the final micelle size and thus the capsule dimension. Separately, TEOS is dissolved in a water medium, e.g., a water-ethanol solution, and the pH of the solution is generally lowered to favor the hydrolysis reaction. Once the hydrolysis is complete, the precursor solution is added to the PCM emulsion; here, a controlled condensation reaction of the precursor happens around the PCM droplets, under basic conditions. The result of the condensation reaction is the formation of an extended silica network around the PCM droplet. The main advantage of sol-gel techniques is the possibility to form a ceramic (e.g., SiO_2, TiO_2, $CaCO_3$) shell, which is generally stronger and stiffer and exhibits a higher thermal conductivity than the polymeric shells [40]. On the other hand, pure metal oxides are usually brittle and subjected to cracks. To mitigate their fragile behavior, the molecular precursor can be chosen that contains a hydrocarbon side group, which will be present also in the final network, resulting in a hybrid organo-ceramic material. For example, organosilica shells produced from methyltriethoxysilane $CH_3Si(OCH_3)_3$ (MTES) are less brittle and more flexible than those produced from TEOS, thanks to the side methyl group that remains in the resulting network [41, 42].

Tab. 2.5: Advantages and disadvantages of the main microencapsulation methods ([32, 43] [34]).

Technique	Advantages	Disadvantages	Particle size (μm)
Physical mechanical methods			
Spray drying	– Low cost – High production yield – Widely available equipment and know-how – Versatile – Easy to scale-up	– Particle agglomeration – High temperature – Uncoated particles – Difficult control of the particle size	5–5,000
Solvent evaporation	Low cost	Lab-scale production	0.5–1,500
Centrifugal extrusion	Suitable for bio-encapsulation	– High temperature – Clogging problems	5–1,500
Physical chemical methods			

Tab. 2.5 (continued)

Technique	Advantages	Disadvantages	Particle size (μm)
Coacervation	– Versatile – High encapsulation efficiency and mild processing conditions – Efficient control of the particle size – Insensitive to water-soluble additives – wide pH working range – Low operating temperature	– Difficult to scale-up – Expensive – Agglomeration – Residual solvents – Not suitable for producing very small (<5 μm) microspheres	2–1,200
Sol-gel	Inorganic shell with high thermal conductivity	– Still under research – Difficult to scale up – Complex reactions involved – Uncoated particles	0.2–20
Chemical methods			
Interfacial polymerization	– Easy to control parameters – Simple and reliable process – High active loading and tunable delivery processes – Versatile and stable mechanical and chemical properties – System temperature up to 80 °C – Relatively low cost – Conducive to scale-up	– Moderate cost – Solvent handling – Non-biocompatible carrier material – Organic solvents	0.5–1,000
Suspension polymerization	– Easy to control parameters – Good heat control of the polymerization reaction	– Moderate cost – Solvent handling	2–4,000
Emulsion polymerization	– Easy control parameters – High MW polymer – Fast process	– Moderate cost – Solvent handling – Impurities from the surfactant	0.05–5

2.4.2 Shape-stabilization

In the field of phase change materials, the term "shape-stabilization" is sometimes used as a synonym for "confinement," also including the microencapsulation techniques [44]. However, most of the dedicated literature refers to "shape-stabilization" to indicate all methods to prevent PCM leakage *besides* encapsulation, and this is the meaning that this term assumes also in this book. Shape-stabilization techniques in-

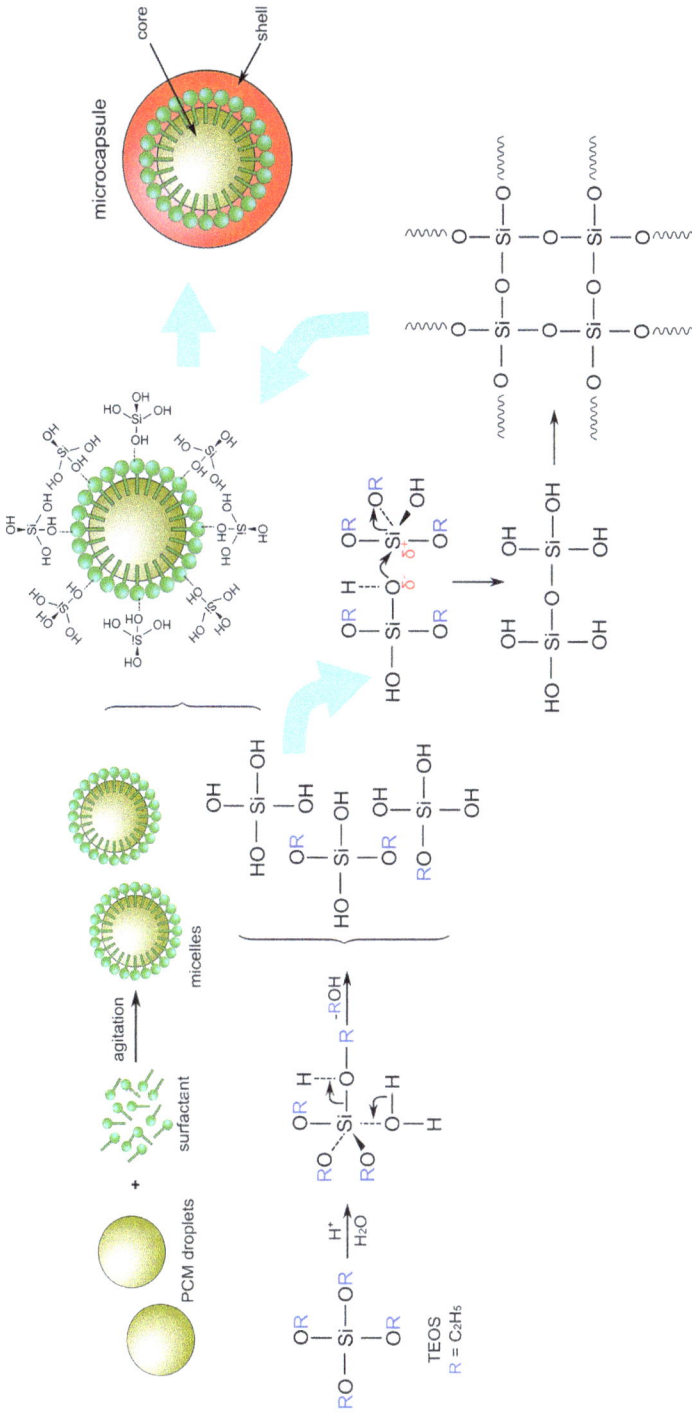

Fig. 2.2: Encapsulation of an organic PCM via a sol-gel process starting from TEOS (duplicated from [5]).

volve the production of a composite PCM via the addition of layered or porous materials, inorganic nanofillers, or polymer matrices, to produce a compound without any manifest leakage or exudation even when the PCM is in the melted state. Such techniques are generally less expensive than microencapsulation and yield a higher thermal conductivity, but the PCM is not completely isolated from the external environment and some leakage may occur after several thermal cycles.

One of the simplest techniques to obtain a composite PCM is the dispersion of nanoparticles, such as carbon nanotubes (CNTs) [45], graphene oxide (GO) [46], expanded graphite (EG) [47, 48], expanded graphite nanoplatelets (xGnPs) [49], nanoclays [50], metallic and metal oxide/nitride nanofillers [51], which increase the mixture viscosity and prevent the leakage thanks to their high specific surface area [44]. The shape-stabilization effect can be achieved by simple melt blending, vacuum impregnation or grafting of the PCM chains onto the nanofillers. Moreover, the inclusion of carbon- or metal-based fillers can enhance the thermal conductivity, thereby improving the energy storage rate and efficiency, especially of the organic PCMs [52].

An analogous effect of enhanced thermal conductivity can be achieved by shape-stabilizing the PCM through a highly conductive foam, where the PCM can be accommodated within the interconnected porosity. Such foams are generally metallic, ceramic, or carbon-based and are characterized by high porosity, interesting mechanical properties, and thermochemical stability. For example, Huang et al. [53] explored the possibility of infiltrating nickel and copper foams with different pore sizes (40, 70, and 90 pores per inch, PPI) with myristyl alcohol, an organic PCM with the chemical formula $C_{14}H_{30}O$, a melting point of 40.4 °C, and a latent heat of fusion of 218.4 J/g, for solar thermal energy storage. The infiltration of molten PCM, performed under vacuum conditions, resulted in a total PCM content of 60–80 wt%, and it increased with the pore size. The thermal conductivity of the impregnated foams was up to 7.5 times higher than that of the neat myristyl alcohol and increased with decreasing pore size (Figure 2.3a).

In another work, Luo et al. [54] shape-stabilized polyethylene glycol (PEG) in a highly aligned N-doped mesoporous carbon (HANC) aerogel matrix. The HANC aerogel was synthesized using a directional freeze-drying method, resulting in a highly ordered structure with aligned carbon sheets. This structure, combined with N-doping and mesoporous characteristics, provided shape stabilization for the PEG through capillary forces and hydrogen bonding with pyridinic nitrogen (Figure 2.3b). The aligned structure also created fast heat transfer pathways, significantly enhancing the thermal conductivity of the composite by 1,500% compared to pristine PEG. Additionally, the N-doping, particularly the pyridinic nitrogen content, improved the latent heat storage capacity and shape stability of the composite. The resulting PEG/HANC composite exhibited excellent leakage resistance up to 80 °C, high latent heat (140 J/g), and enhanced thermal conductivity (4.50 W/(m · K)). This multifunctional improvement in thermal properties made the composite highly effective for battery thermal management, decreasing battery pack temperature by 13 °C and improving discharge capacity by 28%.

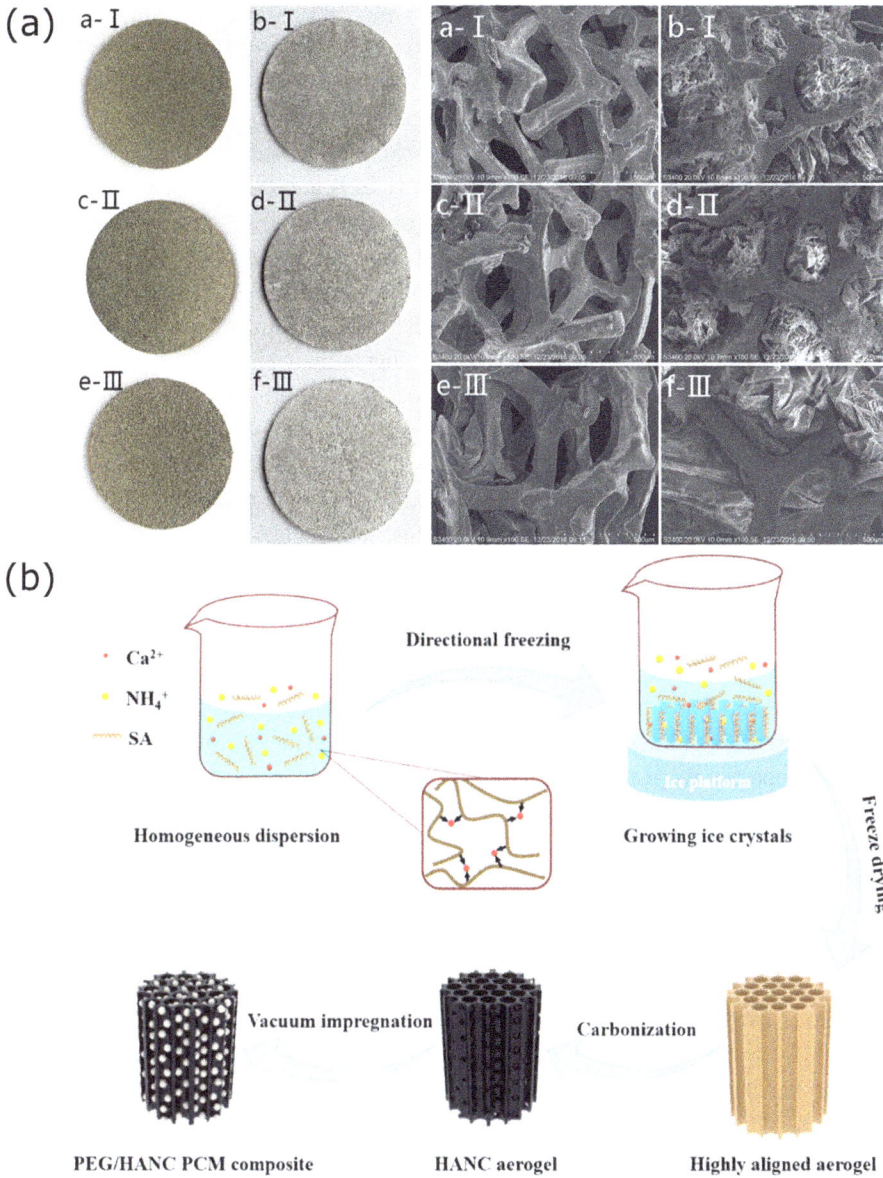

Fig. 2.3: Examples of PCMs shape-stabilized with highly conductive porous materials. (a) Pictures and SEM micrographs of nickel foams before and after impregnation with myristyl alcohol (duplicated with permission from [53]); (b) Method of preparation of a highly aligned N-doped mesoporous carbon (HANC) aerogel matrix and impregnation with PEG (duplicated with permission from [54]).

If the conductive foam also exhibits high mechanical properties, the shape-stabilizing agent can contribute to the mechanical performance of the whole composite, thereby shifting the multifunctionality from the composite level down to the level of the single phase. As will be better described in Chapter 5, this may be beneficial for the overall multifunctionality of the structural TES composite.

Blending organic PCMs with polymer matrices is another diffused technique to prevent leakage in the temperature interval between the transition temperature of the PCM and the melting point (for semicrystalline polymers) or the glass transition temperature (for amorphous polymers) of the surrounding matrix [19]. One of the most widely used strategies consists of combining paraffinic PCMs with polyolefins like polyethylene (PE) and polypropylene (PP), due to their physical-mechanical properties and the chemical compatibility with paraffins, but other commonly used polymers are acrylics, poly(vinyl chloride), polyurethanes and elastomers as ethylene-propylene diene monomer (EPDM) rubbers [19]. For example, Resch-Fauster and Feuchter [55] compared the properties of PE, PP, and an ethylene-propylene copolymer (E/P) as shape-stabilizing matrices for high-molecular weight paraffin waxes. Samples were prepared via melt blending and compression molding; the total amount of PCM in the blends varied between 25 and 38 wt%, which resulted in a phase change enthalpy of 53–71 J/g. A co-crystallization was evident in the PE-PCM composites, while for PP and E/P the crystallization of the PCM was not affected by compounding, but the PCM acted as a plasticizer and a nucleating agent for the surrounding matrix, thereby increasing the crystallinity of the matrix. Nevertheless, the satisfactory heat storage properties were accompanied by a decreasing in the mechanical performance after PCM addition. This was noticed especially above the PCM melting transition, associated to a decrease up to 85% in the elastic modulus and 54% in the tensile strength, which led the authors to conclude that such PCM-polymer blends are not suitable as load-bearing materials.

2.5 Cost and sustainability of PCMs

TES systems based on PCMs are generally more expensive than sensible heat storage materials. Although PCMs are widely studied in many different fields, and new materials with a high latent heat of phase change are constantly being discovered and synthesized, PCMs are currently being massively employed only in building and construction applications [27] and, even in this sector, the time required to recover the initial investment cost can be of up to several decades. Besides the buildings field, PCMs are yet to have a fully developed market or high demand, which determines a considerable cost increase. The majority of the total cost of a PCM system is represented by the cost of the material (some tens of dollars per kWh) and its confinement (encapsulation or shape-stabilization), as the cost of macroencapsulation can reach approximately 20% of the total, while the price of microencapsulation can also represent 50% [27].

For the PCM classes, commercial paraffins are by-products of oil refining and their cost is limited, but it strongly depends on the purity of the material. The price per kilogram ranges from units of dollars for technical grade alkanes (purity ≈ 90%) to tens of dollars for the analytical grade variations with ultra-low concentrations of impurities [26].

The environmental impact of materials and systems must be evaluated throughout the whole life of the component. One of the most complete and valuable approaches to comprehensively assess the environmental footprint of any product or system is the life cycle assessment (LCA) [56], which considers all the inputs and outputs of a system from the extraction of the raw materials to the manufacture, service time, maintenance, and end of life. The inputs are evaluated as the raw materials and the required energy for each step, while the outputs are regarded as all the solid, liquid, and gaseous waste materials released into the environment during the whole life cycle. The PCMs, and in general the TES systems, have a positive environmental impact during their operating period, as they reduce the gap between energy demand and availability, help save primary energy resources, and reduce CO_2 emissions. On the other hand, the other life cycle stages, such as production, fabrication, and disposal, can have a negative impact on the environment. To estimate the balance between these contributions and calculate the net environmental impact, the LCA has been proven an effective approach, as it considers all the single contributions and evaluates the performance even of complex systems. In this way, the LCA helps industries and organizations select specific indicators to describe the effect of a system on the environment and to identify opportunities for improving the environmental impact of the evaluated product.

LCA has been performed on PCM systems especially for building applications, following the general guidelines described in the international standards ISO 14040 and ISO 14044, and the results of this investigation have been illustrated in some recent review papers [56, 57]. Several PCM-containing systems (e.g., polyurethane foams containing microencapsulated paraffins or hydrated salts, PCM-enhanced alveolar bricks, and concrete structures incorporating PCMs) were investigated and compared with more traditional counterparts. Generally, though, the solutions including a PCM resulted as more environmentally friendly than the conventional thermally insulating systems [56]. However, as the implementation of this technology is in its infancy, a deeper investigation is required to fully understand the environmental implications of PCMs and to assess their impact compared to their alternatives [27]. Moreover, some authors [57] highlight issues in LCA methodology, including a lack of accurate environmental data for specific PCMs, the frequent omission of maintenance and replacement considerations in assessments, and inadequately defined functional units in embodied impact studies. These limitations potentially compromise the accuracy and comparability of existing LCA studies on PCMs. The authors suggest that future LCAs should incorporate more comprehensive "cradle-to-grave" assessments, include maintenance and replacement factors, and utilize more precisely defined functional units to improve the objectivity and practical relevance of their conclusions.

2.6 Thermal conductivity enhancement of organic PCMs

One of the main disadvantages of organic PCMs is their inherently low thermal conductivity, which reduces the charging/discharging rate and may inhibit the completion of the phase change process in applications with fast heating or cooling, thereby limiting the full exploitation of the TES capability of the storage medium [58]. To overcome this problem and enhance the heat transfer rate, several solutions have been proposed and implemented, which involve the increase in the heat exchange surface and/or the use of highly conductive fillers, containers, and stabilizing agents [59, 60]. Among these techniques, the addition of highly conductive fillers is the most common to specifically tackle the thermal conductivity issue [61]. Several types of fillers are currently employed, which can be classified as carbon-based, metallic, and ceramic [59, 62].

Carbon-based fillers are among the most popular additives because, besides having high thermal conductivity, they feature considerable thermal stability and low density. They are available in several morphologies, such as carbon fibers (CFs) and nanofibers (CNFs), carbon nanotubes (CNTs) and nanospheres (CNSs), graphene oxide (GO), expanded graphite (EG), and expanded graphite nanoplatelets (xGnPs). Sari et al. [48] impregnated the porous lamellar structure of EG with n-docosane, and the produced PCM composite could be considered shape-stabilized for an EG content of 10 wt%. Moreover, the thermal conductivity increased linearly with the EG content, from 0.22 W/(m · K) for the pure docosane to 0.82 W/(m · K) for the composite containing 10 wt% of EG. Hence, while nanoparticle addition often reduces energy storage capacity, the significant enhancement in thermal conductivity can compensate for this decrease by improving heat transfer rates.

Besides the thermal conductivity, the shape, size, aspect ratio, orientation, and dispersion of nanofillers have a great impact on the final thermal properties of the PCM [58]. For example, smaller nanoparticles (1–10 nm) tend to improve thermal properties more effectively than larger ones, though their performance can be influenced by surface and quantum size effects. Several models have also been developed to study the thermal conductivity enhancement for nanofilled PCMs starting from the properties of the starting materials [61]. Recently, Zhang et al. [45] evaluated the performance of an anisotropic ordered CNT array as a shape-stabilizer and thermal conductivity enhancer for a paraffinic PCM. Embedding the CNT array resulted in a decrease in the phase change enthalpy by 15%, but the thermal conductivity increased from 0.235 W/(m · K) of the neat paraffin up to 12.3 W/(m · K) for the composite PCM in the direction of the CNT axis (for the transversal direction, the thermal conductivity was 4.17 W/(m · K)).

The combination of microencapsulation and the use of conductive materials can further enhance thermal conductivity. For example, Lin et al. [40] encapsulated stearic acid in silica shells, and the microcapsules had a higher thermal conductivity than the neat PCM, due to the synergistic effect between the increase in the heat transfer area and the high intrinsic thermal conductivity of the ceramic shell. The authors also modified the prepared microcapsules by grafting GO on the shell surface and noticed a further increase in thermal conductivity.

2.7 Applications of PCMs for heat management

The present subchapter introduces some examples and case studies about the use of PCMs in several applications of thermal energy storage and thermal management, to highlight the variety of potentialities of such materials and provide the instruments for a deeper understanding of their working principles. This subchapter will focus especially on the thermal management applications, as these are the intended applications for the multifunctional composites object of this book.

2.7.1 Thermal management in buildings

The global energy landscape has witnessed a dramatic shift in recent decades, with space heating and cooling emerging as a critical concern. Since 1990, energy consumption for indoor thermal comfort has more than tripled, presenting significant challenges to electricity grids, greenhouse gas emissions, and urban heat management. This trend has been exacerbated by population growth and record-breaking temperature increases, highlighting the urgent need for sustainable temperature management solutions.

Housing and tertiary buildings account for the consumption of more than 40% of the total primary energy and approximately 19% of the overall CO_2 emissions [4, 63]. Considering the residential buildings, among the energy end-uses, space heating and water heating are responsible for the largest portion of total energy consumption, which is 57% in the U.S., 71% in China, and 80% in the E.U. [63, 64].

The impact of inadequate indoor temperature regulation extends beyond mere discomfort, posing risks to human health and labor productivity for a substantial portion of the world's population. In response to these challenges, initiatives like the Net Zero Emissions by 2050 have proposed comprehensive strategies, including improving building efficiency and promoting passive cooling solutions. The scale of the issue is underscored by the International Energy Agency (IEA)'s report, which revealed that energy demand in buildings reached a staggering 133 exajoules in 2022. As the global community grapples with the dual imperative of ensuring thermal comfort and mitigating environmental impact, one of the most interesting and viable solutions implemented involves the use of PCMs in passive or active energy storage systems, as illustrated in several recent reviews on the topic [65–68].

2.7.1.1 Passive storage systems

Passive storage systems include heating and cooling technologies without an active mechanical device and with little or no external energy input. An example of a passive storage system is represented by the inclusion of PCMs in wallboards, ceilings, or flooring materials, which can store excess energy during the day (peak hours) and release it during the night (off-peak hours), helping to regulate the temperature under extreme weather conditions [19]. As indoor thermal comfort is generally considered to be achieved in a temperature interval between 18 °C and 25 °C, this is also the range of phase change temperatures of the selected PCMs [69].

Although the concept of using LH-TES in buildings has been known for decades, one of the first systematic studies was conducted in 1996 by Feldman and Banu [70], who fabricated PCM-enhanced lab-scale gypsum wallboard samples. The PCM phase, represented by a mixture of fatty acids, was introduced by impregnation (without further encapsulation or shape-stabilization) and accounted for approximately 25% of the total wallboard weight. The thermal storage capacity of the wallboards was studied at different scales. First, differential scanning calorimetry (DSC) was employed to measure the specific melting enthalpy of small specimens and evaluate the uniformity of the PCM distribution. The authors then upscaled the experiments and evaluated the total TES performance of a room lined with these wallboards. They concluded that after the heating/cooling system was shot off, the room temperature could be maintained in the thermal comfort range for several hours longer compared to a room with traditional wallboards, without impairing the air quality [71].

Technologies that allow the storage of excessive solar thermal energy are particularly attractive for releasing heat during the night or to reducing overheating due to solar radiation during peak hours. An example of this second case is provided by Wang and Zhao [72], who proposed a PCM-enhanced curtain to reduce the solar heat gain through windows and thus the energy required for cooling, which is especially useful for modern glass-wall buildings. A numerical investigation demonstrated that the selection of the PCM with the most appropriate melting temperature plays a key role in determining the curtain performance and that the heat gain of the indoor space can be reduced by up to 16.2% with a PCM layer of 5 mm.

There are several ways to integrate PCMs for structural or rigid building elements. In addition to the aforementioned direct impregnation, which can involve exudation of PCM and loss of performance over time, other techniques follow two approaches: the use of microencapsulated PCMs to be mixed with construction materials and the addition of a macroencapsulated or variously stabilized PCM as a supplementary layer [73]. One of the most recent studies implementing the first approach is that of Bao et al. [74], who developed a high performance PCM-enhanced cement composite for passive solar buildings. The PCM phase is represented by a paraffin wax (T_m = 28 °C) microencapsulated in polymeric shells containing graphite flakes, added during microcapsule synthesis to enhance the thermal conductivity. The microencapsulated PCM was mixed with the cement matrix together with nanosilica and short carbon fibers to preserve the mechanical properties and further enhance the thermal conductivity.

2.7.1.2 Active storage systems

Active storage systems are used to store the heat produced when the primary energy source is more abundant or less expensive. In buildings, active storage systems based on PCMs are used to store heat produced by heating systems during the night, so that the energy peak is effectively reduced and shifted to nighttime when the cost of electricity is lower. Lin et al. [75] developed a floor heating technology integrated with shape-stabilized PCM panels. Such panels, made of paraffin with a melting temperature of 52 °C (75 wt%) and polyethylene as a supporting material, were placed under a wooden floor onto electric heaters. Large-scale experiments proved that the system was effective in increasing the indoor temperature and maintaining the temperature within an acceptable range, long after the heaters were switched off.

2.7.2 Smart textiles

The embedding of PCMs in textile fabrics leads to the production of smart textiles that help regulate body temperature and are particularly useful in situations of extreme weather conditions [76–78]. One of the first examples of PCM-enhanced textiles was produced by NASA; nonadecane was added to garment fabrics (e.g., in space suits) to limit the impact of the extreme temperature changes to which astronauts are subjected during space missions [79]. Later, smart thermoregulating textiles were employed to enhance the thermal comfort of mountain outdoor clothing and apparel, as well as blankets, mattresses, and pillow cases [19]. Figure 2.4 illustrates the main ways to introduce a PCM in fibers, yarns, or textiles and the basic working principle of PCM-enhanced apparel.

There are five main ways to embed a PCM in a synthetic textile: (a) the mixing of the PCM with the melt/wet spun polymer in the form of a core filament; (b) the production of core-sheath fibers, in which the core is composed of the PCM and the shell is the supporting polymer; (c) the introduction of PCM microcapsules in the melt/wet spun polymer; (d) the application of microencapsulated PCMs on fabrics using suitable binders or coating materials; and (e) the introduction of a PCM-enriched inner layer (e.g., polyurethane foam containing PCM microcapsules) [80]. Among these techniques, the first three methods are often preferred as they result in versatile multifunctional polymer fibers with the PCM phase stably confined into the surrounding polymer, which reduces the risk of removal during washing [81]. However, the shear

(a)   (b)

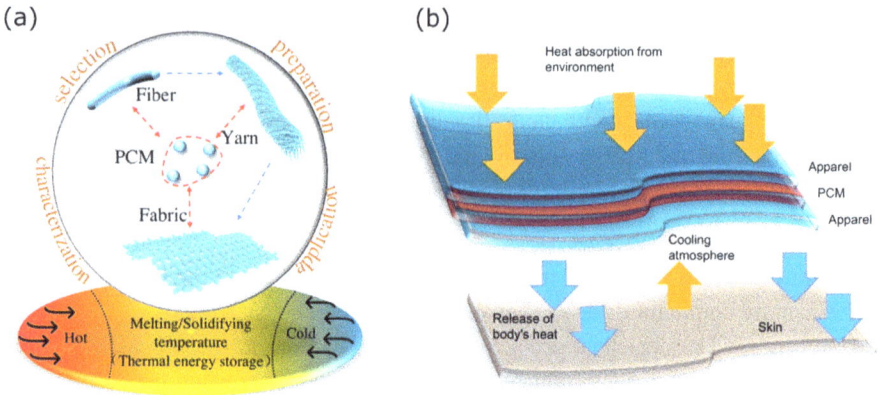

Fig. 2.4: (a) Ways of introducing PCMs in fibers, yarns, and textiles (duplicated with permission from [76]); (b) Working principle of PCM-enhanced apparel (duplicated with permission from [77]).

stresses, high temperature and/or aggressive solvents present during the fiber spinning process could damage the PCM microcapsules, thereby causing PCM leakage and low final phase change enthalpy.

Among the most famous examples of PCM-containing polymer fibers are those marketed by Outlast Technologies GmbH, which produces viscose/rayon and acrylic fibers containing a microencapsulated PCM with a total phase change enthalpy of 1–20 J/g [82]. The same company also produces PCM-enriched coatings for outwear, footwear, bedding and seating, and patented a technology to print PCM microcapsules onto flat fabrics, which can be directly introduced into the fabric production line [83].

The embedding of PCMs in polymeric fibers is interesting not only for the production of smart textiles but also for fabricating multifunctional polymer filaments that can be co-woven with continuous reinforcing glass or carbon fibers, to produce a multifunctional yarn containing the matrix, the reinforcement and the PCM. This is an interesting route that can be explored to fabricate thermoplastic composites with TES properties.

2.7.3 Thermal management of electronics

Electronic devices are well known to be sensitive to temperature, as their performance and lifespan depend strictly on their maintenance within a precise temperature range, with a particular attention to avoiding overheating. As electronic components are being equipped with increasingly sophisticated electronics and their dimensions have decreased, the risk of overheating has also increased. Without an appropriate thermal management system, the heat generation and associated temperature rise may deteriorate performance, bring critical components to failure, and decrease user-device interaction comfort [84, 85]. Overheating is among the most common causes of electronic component failure, as approximately 55% of failures can be related to high-temperature problems or poor thermal management [86]. It has been shown that a decrease of 1 °C can decrease the failure rate of up to 4%, and an increase of 10–20 °C can double the failure probability [86].

An effective thermal management system must also comply with the weight and size limitations, as these design parameters are increasingly important for electronic components that must be carried around, such as portable and wearable electronic devices, but also batteries and circuitry for electric vehicles (EVs). From this perspective, PCMs are becoming an attractive alternative to more bulky solutions, such as natural or forced convection (active cooling), because electronic devices do not normally need to operate continuously for long periods [10]. When the device is at a heat peak and its temperature starts rising, the PCM melts and absorbs excess heat, thereby preventing an excessive temperature burst (passive cooling). When the temperature starts to decrease again, the PCM crystallizes and releases heat back to the environment. The ideal PCM for this application has a high energy density per unit mass and volume, a phase change temperature slightly below the maximum operating temperature of the component, and a high thermal conductivity.

Pioneering experimental work on the use of PCMs in mobile electronic devices was carried out by Tan and Tso [87], who assessed the efficacy of a passive cooling unit based on n-eicosane for the thermal management of small handheld personal computers (personal digital assistants). The PCM was contained in an aluminum case and placed under heaters simulating the heat generation units of such a device, that is, the processor and other integrated circuit packaging. The authors concluded that the PCM units were indispensable for maintaining the working temperature of the device under an acceptable threshold of 50 °C and that the efficacy of the heat storage unit depended not only on the amount of PCM but also on its orientation, which determined the heat flux distribution in the entire device. The same concept was developed by Tomizawa et al. [88], who numerically and experimentally investigated a passive cooling unit for mobile phones that contained commercial microencapsulated paraffin with a melting temperature of 32 °C. The PCM was mixed with polyethylene and molded as a sheet for inclusion in the mobile phone. A schematic of the experiment is shown in Fig. 2.5. The authors concluded that the PCM sheet actively contributed to slowing down the temperature increase, and this effect was more evident with thick sheets.

PCMs can also be employed for the thermal management of EV batteries to support or replace traditional cooling systems based on liquid/air circulation [27, 89]. The first attempts to integrate PCMs in the automotive field date back to the early 2000s, when Al Hallaj and Selman [90] proposed a battery pack in which each cylindrical Li-ion cell was wrapped with a PCM layer with a melting temperature in the range of 30–60 °C. The authors experimentally and numerically demonstrated that the total temperature fluctuation was considerably lower with the PCM.

(a)

(b)

14 mm

122 mm

62 mm

Front case #1

Front case #2

PCM (PE) #1

PCM (PE) #2

Heater

Substrate #1

Substrate #2

Rear case #1

Rear case #2

(c)

| Front case (1 mm thick) | PCM (PE) sheet (1 mm or 4 mm thick) | Copper sheet (0.08 mm thick) |

14 mm

Inner air

Base (0.5 mm thick)

Rear case (1 mm thick) — Ceramic heater (1.27 mm thick) — Nut (3 mm high) — Bolt

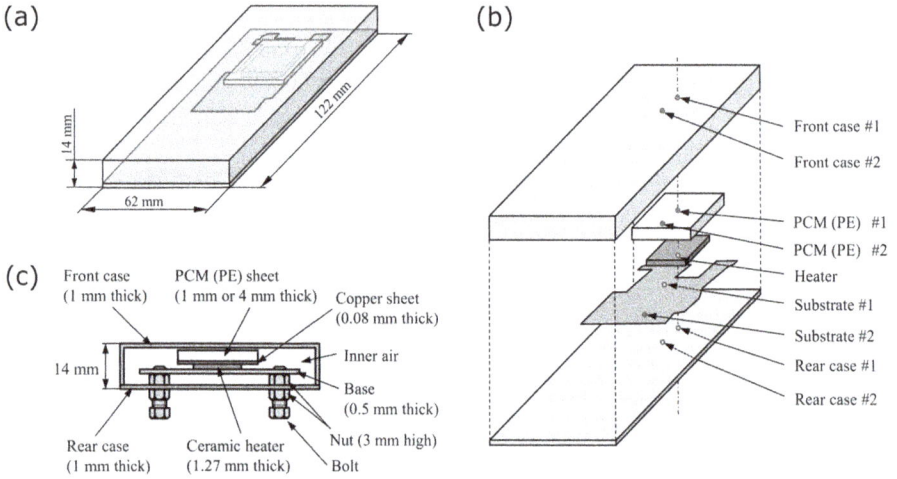

Fig. 2.5: Schematic diagram of the mobile phone heating simulator. The front case is made of polycarbonate boards and the rear case of acrylic board. A ceramic heater for simulating large-scale integration (LSI) is fixed on a substrate by a double-sided thermal tape. (a) General scheme and dimensions; (b) elements and position of the test thermocouples; (c) cross-sectional view (adapted with permission from [88]).

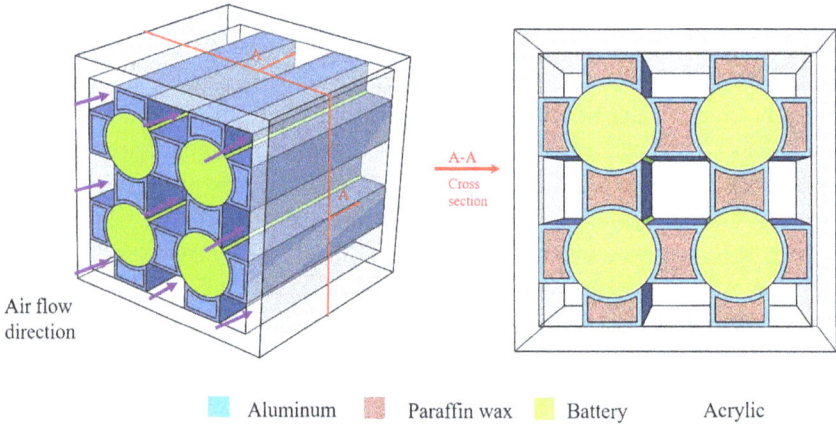

Air flow direction

A-A
Cross section

■ Aluminum   ■ Paraffin wax   ■ Battery   Acrylic

Fig. 2.6: Hybrid thermal management system for EV batteries combining a PCM and forced air circulation (duplicated from [93]).

Successive approaches have attempted to address the low thermal conductivity of PCMs and the need to improve the thermal uniformity inside battery packs. The most promising solutions involve metal or graphite foams as shape stabilizers and thermal conductivity enhancers. Goli et al. [91] prepared a composite PCM combining paraffin wax and exfoliated graphene, which exhibited a thermal conductivity two orders of magnitude greater than that of the neat paraffin. Experimental and numerical simulations proved that this composite PCM led to a considerable decrease in the heating rate and maximum temperature inside the battery pack, outperforming not only the design without a PCM but also that with unfilled paraffin. More recently, Zou et al. [92] studied the introduction of various carbon-based nanofillers (e.g., EG, CNTs, graphene) in a paraffin wax, to produce a shape-stabilized PCM to be employed in a 38,120-type $LiFeO_4$ battery pack. They found that this composite PCM could reduce not only the maximum operating temperature but also the temperature oscillations. Other approaches have tackled the problem of thermal management using hybrid systems that combine PCMs with forced convection systems, thereby exploiting active and passive cooling. Qin et al. [93] included a paraffin wax ($T_m = 56$ °C) in the design of a forced air circulation system, showing that the combination of the two contributions helps in maintaining the temperature within the acceptable range.

2.7.4 Biomedical applications

Owing to their ability to store excess heat at nearly constant temperatures, PCMs are becoming appealing for biomedical applications that require thermal protection. Smart thermoregulating biocompatible fabrics can be used to keep the skin within a certain temperature range, which can be useful for heat/cool therapy and burn wound dressing.

Controlled heat absorption and release can also be employed to protect living tissues from excessive heat and cold. For example, Król et al. [25] added shape-stabilized PEG to a PMMA-based bone cement formulation to decrease the peak polymerization temperature and prevent overheating and damage to surrounding tissues. The results showed that this biodegradable and biocompatible PCM, added in a weight fraction of 15%, led to a significant decrease in the peak temperature, from 70.2 °C to 58.3 °C. Moreover, the PEG addition determined the formation of micrometric pores on the bone cement surface, which caused a decrease in the compression strength, but also favored cell proliferation during bone formation and improved the final osteointegration.

PCMs can also be used during cryosurgery of tumors, as in the concept proposed by Lv et al. [94]. If PCMs are injected into the tissues surrounding the tumor, the heat released during PCM crystallization prevents an excessive temperature drop in the healthy tissue, thereby protecting it from damage and necrosis. Theoretical studies have shown that PCM addition can maximize the necrosis of tumor cells and minimize injury to the surrounding tissues.

2.7.5 Thermoregulating packaging

Food, medical supplies, and other perishable products must be marketed and distributed while kept in a specific temperature range, usually under refrigerated conditions, to avoid spoilage and preserve product quality and safety. This may be an issue during product transportation and temporary storage in provisional warehouses that lack temperature control [4, 95, 96]. The traditional way to address this problem is to provide effective insulation; however, this might not be sufficient, and the tendency to produce low-cost and lightweight packages often leads to packaging materials with limited thermal buffering capacity [95].

A recent conceptual advancement was obtained by imbuing packaging with heat management properties. Organic PCMs with a melting temperature between −5 °C and 10 °C are a suitable choice for producing smart thermoregulating packages that protect the contents from perishing while reducing the need for active refrigeration and energy consumption [96].

An example of PCM-enhanced packaging was proposed by Johnston et al. [97], who employed a highly porous nanostructured calcium silicate powder to shape-stabilize a paraffin wax with a melting point of 8 °C. The authors used this PCM composite to line a paperboard container, which was able to maintain an inner temperature below 10 °C for 5 h after the outside temperature was increased from 2 °C to 23 °C. Recently, for food applications, there has been a strong tendency towards choosing the PCM in the fatty acid family because they are biodegradable and can be produced from renewable resources. For example, Ünal et al. [98] designed a three-layer cardboard box where the middle layer was filled with bulk or microencapsulated octanoic acid, and the experimental results showed that the PCM-enhanced box was able to provide up to 8.8 h of thermal buffering.

2.8 Flame resistance of organic PCMs

Owing to their chemical composition, organic PCMs, and paraffin waxes in particular, are highly flammable, which is an issue, especially when PCMs are employed as thermal management media in buildings, textiles, and electronic devices.

To address this problem, the most common solution is to incorporate conventional flame retardants (e.g., clays [99], metal oxides [100], intumescent agents [101], and halogenated compounds [102]) in the microcapsule shell or directly in the PCM mass, where they can also contribute to improving the shape-stabilization and thermal conductivity. Zhang et al. [47] incorporated expanded graphite and an intumescent fire retardant in a shape-stabilized polyethylene/paraffin PCM and noticed that the fire resistance increased with the content of fire retardant, as the values of peak heat-release rate measured in cone calorimetry tests were halved with a total filler content of 25 wt%. Similar results were reported by Li et al. [101], who used the same strategy to improve the flame resistance of a polypropylene/paraffin PCM. The authors concluded that a total flame-retardant content of 30 wt% considerably increased the PCM performance in the limiting oxygen index (LOI) and cone calorimetry tests.

3 Polymer-matrix composites

The present chapter provides a concise overview of polymer-matrix composite materials, with a special focus on fiber-reinforced polymer (FRP) composites. After a brief introduction illustrating the definition and classification of polymer composites, the chapter presents the most diffused matrices and reinforcements and describes some of the state-of-the-art fabrication and characterization techniques.

3.1 Introduction to polymer composites

This section describes the general definition and classification of polymer composites and then focuses on their applications.

3.1.1 Definition and classification

Composite materials are generally regarded as those consisting of two or more distinct materials or phases, which exhibit remarkably different mechanical and/or physical properties. Therefore, the properties of the resulting material are noticeably different from those of each constituent, and the composition is tailored to obtain the combination of properties that best suits the constraints given by the production process and the application [103].

Composite materials pervade our world. Besides being the most widespread material type among natural and biological materials, composites have been produced and used by mankind for thousands of years; one of the first examples of man-made structural composites is represented by the straw-reinforced mud bricks found in Mesopotamia, dating back to 4900 BC. However, it is only in the last century, with the advent of high-strength synthetic fibers and the enormous advances in polymer chemistry and technology, that composite materials can be made that offer performance comparable to or even superior to those of well established structural materials such as metals [104].

Composites are generally constituted by a continuous phase, the *matrix*, and one or more discontinuous phases, the *fillers*, which are generally stronger and stiffer than the matrix and therefore can also be called *reinforcements* or *reinforcing agents* [105]. Composite materials can be classified according to the matrix material as polymer matrix composites (PMCs), ceramic matrix composites (CMCs), and metal matrix composites (MMCs). Each of these classes has a particular set of properties and specific application fields, as the matrix has a strong influence on several mechanical properties of the composite, such as transverse modulus and strength and the properties in shear and compression, as well as on the maximum service temperature [103].

https://doi.org/10.1515/9783111111865-003

For the purpose of this book, only polymer composites are taken into account, as they are the most widely used in structural and semi-structural applications at low-medium temperatures (0–250 °C), due to their lightness and high specific stiffness and strength.

Polymer composites combine a polymer matrix with one or more fillers, commonly added to improve stiffness, strength, and high-temperature performance. As the mechanism of improving a property strongly depends on the filler geometry, it is convenient to classify polymer composites according to the reinforcement type, shape, and size. The reinforcement can be made of fibers or particles, as observable from the commonly accepted classification reported in Fig. 3.1. The most common types of reinforcements and matrices used to produce polymer composites are described in Sections 3.2 and 3.3, respectively.

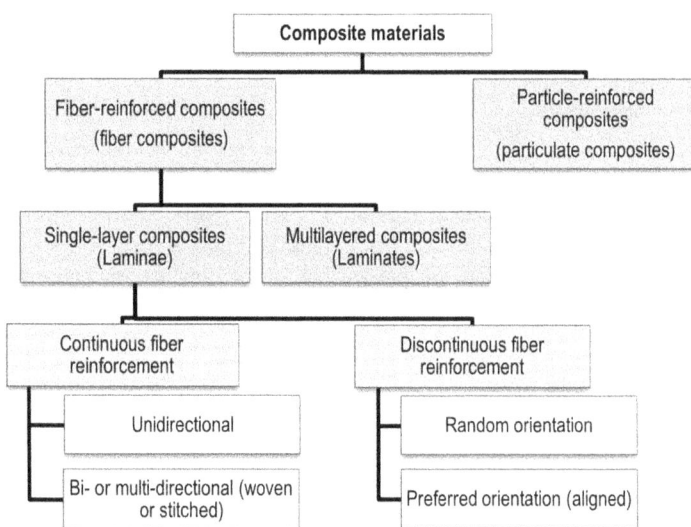

Fig. 3.1: Classification of composite materials according to the geometry of the reinforcement (data from [103]).

3.1.2 Significance and applications

The main advantages of structural polymer-matrix composites are (i) the possibility of being tailored for optimum strength and stiffness in different loading directions, and (ii) the combination of low density and high strength and modulus. This latter aspect results in higher specific (i.e., normalized by density) mechanical properties than those of comparable aerospace metal alloys, which in turn allows the design of light-weight structures, thereby leading to improved performance and fuel savings, especially in the automotive and aerospace industries. Other advantages of structural composites compared to lightweight metal alloys are the high fatigue life and corro-sion resistance [104]. Due to these features, composite materials are mainly applied in construction and transportation fields, but their application is also expanding in ma-rine structures, sports equipment, and infrastructures.

On the other hand, among the main disadvantages of composite materials are the expense of raw materials, fabrication, and assembly, as well as the higher sensitivity to temperature and moisture, which generally lead to a decrease in performance. Moreover, composites exhibit poor strength in the out-of-plane direction where the matrix carries the primary load, but this aspect can be mitigated by orienting the re-inforcement properly. Composites are also more susceptible than metals to impact damage, are prone to suffer from delamination, and can be challenging to repair.

3.2 Reinforcements

Reinforcements in polymer composites are mainly divided into fibers and particles [105]. For most of the structural and semi-structural polymer composites, the reinforc-ing agent is almost exclusively constituted by fibers, because their stiffness and strength combined with the high aspect ratio make them the most suitable reinforce-ment for structural applications. The following subchapters will focus on the main fiber classes and give some hints on the other reinforcement types.

3.2.1 Continuous and discontinuous fibers

Fibers are the primary load-bearing components in FRP composites. Several types of fibers are available for the reinforcement of different matrix materials in different applications, and the most widely used are glass, carbon, and aramid fibers.

Glass fibers dominate the market for large structures such as wind turbine blades, ships, and civil engineering structures, due to their low cost, high tensile strength, high impact strength, and good thermal and chemical resistance. Although there are many kinds of glass fibers, the three most diffused as reinforcement are E-glass, S-2 glass, and quartz fibers. E-glass fibers are the least expensive and the most common, exhibiting a good combination of tensile strength (3.5 GPa) and modulus (70 GPa). S-2 glass is more expensive but stronger than E-glass, featuring a tensile strength of 4.5 GPa and a modulus of 87 GPa, and it has better mechanical performance at elevated temperatures. Quartz fibers are made of a rather expensive, highly pure sil-ica glass, used primarily in demanding electrical applications [103].

Carbon and graphite fibers are the most prevalent fiber types used in highly de-manding applications such as aerospace, but their use is also increasing in the general automotive field. They outperform glass fibers in tensile stiffness and fatigue resistance. Graphite fibers are subjected to heat treatments of graphitization above 2,000 °C, which determines a carbon content higher than 99% and the growth of bigger and more aligned crystallites. This is the reason for their superior elastic modulus, greater than 345 GPa. On the other hand, carbon fibers do not undergo graphitization and have lower carbon contents (93–95%) and lower stiffness [103, 106].

Aramid fibers (e.g., Kevlar) are the preferred choice when good impact energy-absorbing properties are required, such as in the military field.

An overview of the properties of these fiber types is reported in Tab. 3.1.

Tab. 3.1: Typical properties of common reinforcing fibers (data from [104, 107, 108]).

Fiber	Density (g/cm^3)	Young's modulus (GPa)	Tensile strength (MPa)	Strain at break (%)	Coefficient of thermal expansion (10^{-6}/K)	Common filament diameter (µm)
E-glass	2.58	72.4	2000–3,450	2.6–4.8	4.9	5–20
Kevlar	1.45	130 (axial); 10 (radial)	3,000	2.3–4	−6/−2	12
HM carbon	1.95	380 (axial); 12 (radial)	2,400	0.6–1	−0.75	5–10
HS carbon	1.75	230 (axial); 20 (radial)	3,400	1.3–2	−0.4/−0.6	5–10

HM = high modulus; HS = high strength.

Fibers are available on the market in several forms to suit different composite fabrication processes. Single filaments are produced with diameters of units to tens of micrometers, and they are collected in single bundles called strands or tows. Such bundles can be gathered in rovings, which generally contain 20 to 60 bundles, to produce the desired linear density. The strands can also be twisted together to form yarns. The most effective way to tailor the architecture of a composite so that the directional dependence of strength and stiffness matches that of the loading environment is to weave tows, rovings, and yarns into fabrics, which can be unidirectional or bidirectional and have a different weave, classified by the pattern of interlacing (e.g., plain, twill, satin weave, see Fig. 3.2) [104, 108].

Unidirectional fibers or woven textiles can also be impregnated with a controlled amount of resin to form a prepreg, an important product form largely used for advanced composite manufacturing. Prepregs usually consist of single textile layers impregnated in B-staged (advanced to a tacky semisolid) resin, which are kept refrigerated to prevent further advancement and change in the resin state [109].

Discontinuous fibers are obtained by chopping rovings into small fiber lengths, which can vary between 1 and 50 mm, while even shorter fibers can be produced by

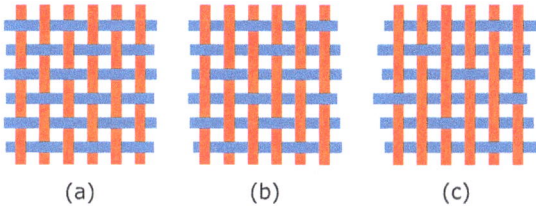

Fig. 3.2: Common two-dimensional weave styles: (a) plain; (b) twill; (c) satin (adapted from [104]).

hammer-milling. Long chopped fibers are usually integrated into thermosetting resins for compression and transfer moldings, while shorter chopped reinforcements are suitable to be blended with thermoplastic systems for injection molding. Milled fibers provide a good combination of reinforcing properties with processing ease and are used to reinforce thermoplastics when the strength requirements are low to moderate.

3.2.2 Particles and other reinforcements

Particle-reinforced composites, or particulate composites, are not generally employed for structural applications. Due to the much lower aspect ratio, particles are less effective than fibers in enhancing mechanical properties such as stiffness and strength of the host polymer. On the other hand, particulate composites are widely employed where the demands on mechanical properties must be subordinated to cost and/or processing requirements. In this sense, particulate composites fill the properties and applications gap between fiber-reinforced composites and unreinforced plastics [104, 110].

Particulate reinforcements include a wide variety of materials, which can be classified as natural or synthetic. Natural reinforcing fillers are of mineral origin and are mainly represented by carbonates (e.g., calcite and dolomite), clays (e.g., aluminum silicates), and talcs (e.g., magnesium silicates). Such fillers are used for cost reduction, but they also increase the stiffness and strength of the host polymer; some mineral fillers are used for their good electrical insulation properties, chemical inertness, low water absorption, and low coefficient of thermal expansion [110].

Synthetic reinforcing fillers include carbon black, synthetic silicas, hydroxides, basic carbonates, and precipitated calcium carbonate. Carbon black is produced on a large scale for being incorporated in polymer matrices, and it is especially used as a reinforcing agent for elastomers in tire applications, as it improves tensile strength, tear strength, stiffness, abrasion resistance and dynamic-mechanical properties of the resulting composite [111]. It is also used in some thermoplastics to improve the resistance to outdoor conditions (UV protection) and in the production of antistatic compounds [112].

Among the most important properties of reinforcing particles are geometrical factors like shape and size. Particle shape is fundamental in determining the final composite stiffness and strength, as well as the melt rheology and the surface smoothness [111]. The shape is determined not only by the chemistry and crystal structure of the filler but also by the production conditions. The particles can be isodimensional (roughly spherical or irregular, 0D) or can assume a preferred orientation if they are acicular (1D) or platy (2D). In the latter case, they can be oriented during processing, which may produce an anisotropic or orthotropic composite. Some common particle shapes can be observed in Fig. 3.3. Despite being such an important parameter, defining a particle shape is not always easy, as the literature abounds with vague terms and it can be difficult to distinguish between primary particles, aggregates, and agglomerates [113].

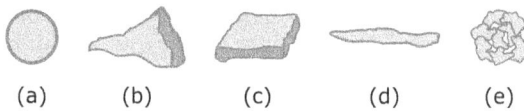

(a)   (b)   (c)   (d)   (e)

Fig. 3.3: Common particle shapes. (a) spherical; (b) irregular; (c) platy; (d) acicular; (e) porous aggregate (adapted from [110]).

Another fundamental parameter is the particle size. Size is determined by the processing conditions, affects a wide variety of properties, and is fundamental for determining the specific surface area (SSA) of the filler, which influences the extent of the filler-matrix interface and all the related factors. The size of the particulate fillers in polymer composites is micrometric and ranges from units to hundreds of micrometers. Since size is such an important property, a variety of techniques have been developed to measure it, such as sieving, sedimentation, laser diffraction, or microscopic techniques. The SSA is usually measured by the permeability of a packed particle bed to a fluid or by the quantitative absorption of nitrogen according to the Brunauer, Emmett, and Teller (BET) method. The value of particle size can be strongly influenced by the measurement technique and complicated if the particle shape is irregular. Moreover, besides the average particle size, it is also important to consider the particle size distribution, which can be broadened also by the presence of particle fragments, aggregates, and agglomerates [105].

A filler class that is gaining increasing attention is that of nanofillers, which include particles having at least one dimension in the nanoscale. To make this definition less vague, some authors [110] have proposed including in this class only the fillers that, when dispersed in a polymer matrix, are made of effective particles with at least one dimension below 20 nm. In this way, the SSA is at least 150 m^2/g, being one or two orders of magnitude higher than that of microfillers. This definition excludes fillers such as carbon black: even though the primary particles meet the conditions, they are strongly aggregated into larger structures that can be considered the effective particles.

Due to their considerable SSA, the surface is remarkably important for nanofillers. Even though they are used in lower volume fractions than microfillers, the total filler-matrix interface is still significantly higher. This determines an increase in the polymer fraction that is modified by the interaction with the filler surface and may have microstructure and properties remarkably different from the bulk polymer. The most common nanofillers include intercalated nanoclays, carbon nanotubes and nanofibers, and colloidal or fumed silica. Although some nanofillers are added to increase the mechanical properties of the host polymer, such as stiffness and heat deflection temperature (HDT), nanofillers are mostly employed to enhance functional properties, e.g., to reduce the gas and fluid permeability or to increase the flame resistance or the thermal conductivity [114, 115].

3.3 Matrices

In a composite, the matrix binds the reinforcement together, transfers the load to the fibers or particles, and protects the fillers against the surrounding environment and possible damage due to handling. The matrix is also critical in compression loading, to prevent premature failure due to fiber buckling, and it supplies the composite with toughness and damage tolerance. Polymer matrices are the most widely used for fiber composites due to their low cost, easy processability, good chemical resistance, and low density. On the other hand, their low thermal resistance generally limits their employment to applications below 300 °C [113].

Polymer matrices for advanced composites are generally classified as thermoplastics or thermosets. Thermoplastics are high-molecular-weight polymers with no intermolecular covalent bonds; this requires them to be processed in the molten state, some degrees above the melting temperature or tens of degrees above the glass transition temperature, according to their semi-crystalline or amorphous nature. After being molded, their shape is consolidated upon cooling. Conversely, thermosetting polymers are supplied as monomers or oligomers having low molecular weight and low viscosity, and they are converted into a three-dimensional crosslinked structure upon heating. This allows them to be processed at mild temperature and pressure conditions, but the crosslinking (curing) step, which traditionally occurs in an autoclave or an oven, can be time-consuming and limits high output production. The result of the crosslinking is a hard, non-crystalline, and often transparent resin with high modulus and strength, which is used for high-end applications in combination with high-performance fibers. Unlike thermoplastics, thermosetting-based composites cannot be post-thermoformed, reshaped, or welded, which limits their repairability and recyclability [104].

The following two sections report examples, advantages, and applications of each of these two polymer classes.

3.3.1 Thermosetting matrices

Thermosetting matrices encompass epoxies, polyesters, vinyl esters, bismaleimides, cyanate esters, polyimides, and phenolic resins. In the low-medium temperature range (up to 135 °C), polyesters and vinyl esters are extensively used for commercial applications, while epoxies are the preferred choice for high-performance composites and adhesives due to their better mechanical properties and higher environmental resistance. Bismaleimides and phenolics are the most suitable resins in the medium-high temperature range (up to 175 °C), while polyimides are the matrix of choice for applications at very high temperatures, up to 315 °C [106].

Epoxy matrices are extensively used for high-end applications due to their strength and excellent adhesion. They also exhibit low shrinkage during curing and high processing versatility. Epoxy resins are supplied as a bi-component system, as they are obtained through the chemical reaction between a prepolymer, carrying at least two oxirane rings, and the curing agent, usually containing amine groups. The reaction of oxirane ring opening and crosslinking with the curing agent is reported in Fig. 3.4.

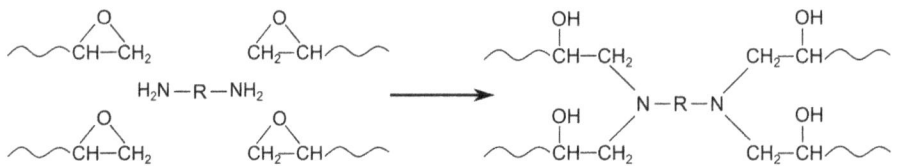

Fig. 3.4: Example of oxirane ring opening and crosslinking reaction with an amine curing agent (adapted from [104]).

Commercial epoxy matrices generally comprise one major and two minor epoxy prepolymers and one or two curing agents. The minor epoxies are added to tune the viscosity and/or to improve temperature resistance, toughness, and moisture absorption. The most widely used major epoxies, especially in the aerospace sector, are bisphenol A diglycidyl ether (DGEBA) and tetraglycidyl methylene dianiline (TGMDA). The epoxy formulation also contains diluents in small amounts (3–5%), to reduce the curing shrinkage, increase the shelf and pot life, and control the exothermicity of the curing reaction. Typical diluents include butyl glycidyl ether, cresyl glycidyl ether, phenyl glycidyl ether, and aliphatic alcohol glycidyl ethers [104].

During the curing of an epoxy resin, higher temperatures and longer times determine a higher crosslink density and, therefore, a higher glass transition temperature, especially if the precursors have high functionality (e.g., four reactive end groups per molecule). To avoid excessive brittleness, the resin can contain toughening agents, but this may result in lower thermal resistance [104].

3.3.2 Thermoplastic matrices

Thermoplastic composites received considerable attention during the 1980s–1990s, when massive investments were made to try to substitute thermosets with thermoplastics in the fabrication of aerospace composites. However, although thermoplastics have some undeniable advantages, their drawbacks limit their diffusion in some high-end fields, and therefore the vast majority of applications involving high-performance continuous fiber composites still prefer thermosetting matrices [104].

As thermoplastics are not crosslinked, they are generally tougher than thermosets, and therefore they have attracted the attention of the aerospace industry as they are more damage-tolerant and have greater low-velocity impact resistance. However, the toughening mechanisms available today for thermosets yield a toughness comparable to that of thermoplastic systems. Moreover, thermoplastics exhibit lower moisture absorption than thermosets, but some thermoplastics, especially those having an amorphous structure, also show low solvent resistance [112].

Another potential advantage of thermoplastics is that they are fully reacted, and therefore they present a low risk of chemical hazard for the workers, who are not exposed to low-molecular-weight components. Additionally, they do not require refrigeration and have an infinite shelf life. Their state of fully reacted polymers also makes their processing, in principle, easier and faster. The shaping and consolidation time for thermoplastics is shorter than that of the thermosets, as it takes minutes instead of hours. Nevertheless, thermoplastics require high temperatures, in the range of 250–450 °C, considerably higher than the curing temperatures for thermosets (120–175 °C), which demands presses and tooling materials that can withstand this temperature regime [112].

Due to the possibility of being remelted, thermoplastics can be joined via techniques such as resistance welding or ultrasonic welding. They can also be post-thermoformed, which is very attractive as it suggests the possibility of producing continuous-fiber-reinforced flat boards to be subsequently cut and thermoformed into the desired shape. However, since traditional continuous fibers have very little extensibility, this can be achieved only with very simple geometries, and also defect healing via remelting can hardly be practiced without fiber distortion [103].

Due to these difficulties, the market for continuous fiber thermoplastics is limited to some specific polymers and applications. The most important matrices in this field are high-performance thermoplastics like polyetheretherketone (PEEK), polyetherketoneketone (PEKK), polyphenylene sulfide (PPS), and polyetherimide (PEI). They are highly aromatic thermoplastics with high glass transition temperatures, good mechanical properties, and good flame resistance [112].

Since the scarce diffusion of continuous fiber thermoplastic composites is mainly linked to the difficult processability of high-performance thermoplastics, new formulations are being developed that are supplied as liquid low-molecular-weight compounds and can be processed at room temperatures with the fabrication techniques typical of thermosets. The most recent example of such materials is the Elium® resin (Arkema, France), an acrylic methyl-methacrylate-based formulation supplied as a low-viscosity liquid and processable at room temperature. The final material has a glass transition temperature of approximately 100 °C, and its thermomechanical properties are similar to those of an epoxy resin, but it preserves the advantages of a thermoplastic composite related to post-thermoformability, weldability, and recyclability [116, 117].

On the other hand, traditional thermoplastics like polypropylene and polyamides are largely employed as matrices for discontinuous fiber composites in semi-structural applications in the automotive, sporting goods, and electronic industries, due to their relatively low cost, easy processing, and superior mechanical properties over unreinforced polymers [118].

3.4 Fabrication of fiber-reinforced composites

Unlike the majority of metals and unreinforced plastics, for which the material is first produced and then processed into the final product, fiber composite parts are generally produced together with the creation of the material. This is more evident for thermosetting composites, in which the final shape and material properties are produced during the matrix crosslinking. In thermoplastic composites, it is more common to fabricate the material first and subsequently shape it, but some composite features, like the fiber length and orientation, can still be influenced in this latter step. The following sections will summarize the most important processing routes for continuous and discontinuous fiber composites.

3.4.1 Continuous fibers and thermosetting matrices

For this class of composites, the fabrication processes can be classified as (i) wet-forming processes and (ii) processes using premixes or prepregs [103]. In the wet-forming processes, the resin is still fluid when the final component is formed and hardens during curing under heating. Such processes encompass hand lay-up, resin-transfer molding, bag molding, filament winding, and pultrusion.

Hand lay-up is the oldest, simplest, and still the most diffused technique for the manufacture of small and large fiber-reinforced products, especially in applications where a low production volume does not justify the high plant size and cost of other fabrication techniques. The typical layup for a bag molding process is illustrated in Fig. 3.5.

The composite laminate is only one of the many layers composing the bag layup. These layers include release agents and films, peel plies, bleeder/breather plies, bagging film, and sealant tape. Release films and agents are used to separate the composite from the mold or the breather/bleeder materials; the release film can be porous to

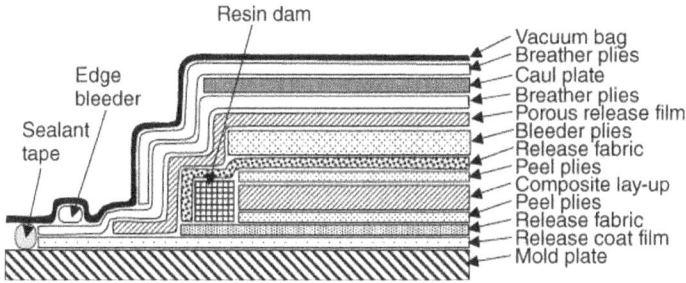

Fig. 3.5: Typical layup for a bag molding process (duplicated with permission from [103]).

allow excess resin to flow through the film. Peel plies protect the surface of the molded part from contamination. Bleeder/breather plies are porous fabrics or nonwovens used to absorb excess resin during processing and to allow air and volatiles out of the composite during curing. Finally, bagging films form a barrier between the composite laminate and the external environment (e.g., oven or autoclave environment) [106].

In the processes using premixes, the material is supplied as an intermediate product, which can be a bulk molding compound (BMC) or a sheet molding compound (SMC) containing fibers and a partially cured matrix. The use of premixes and prepregs simplifies manufacturing, increases the possibility of automation, and helps in obtaining a uniform filler distribution and a higher fiber weight fraction [103]. On the other hand, it limits the *a posteriori* modification of the material composition with additional fillers, fiber coatings, or resin additives.

3.4.2 Continuous fibers and thermoplastic matrices

The high viscosity of polymer melts complicates the incorporation of continuous fibers into thermoplastic matrices, but several techniques have been developed to produce "thermoplastic prepregs" or semi-manufactured thermoplastic composites containing continuous fibers [112]. Unlike the thermosetting prepregs, these have the advantage of having a virtually infinite shelf life and, when required, they can be stacked and consolidated in a laminate by applying heat and pressure.

Among the techniques developed to produce thermoplastic prepregs, hot-melt impregnation is one of the most diffused for semicrystalline polymers. Collimated fiber tows are pulled through a die attached to the end of an extruder, and a thin sheet of polymer melt is deposited on the fibers, previously separated by an air jet. The prepreg is then cooled and wound on a take-up roll. For polymers that can be easily dissolved in a solvent, this process can be substituted by solution impregnation. It produces tacky and drapable prepregs, but accurate solvent removal is a critical issue, fundamental for obtaining high-quality laminates with low porosity [108]. A different technique is film stacking, in which polymer sheets are interleaved between woven fabrics or random fiber mats, and then the layup is heated and pressed to melt the polymer and wet the fibers.

Fiber mixing is an interesting process in which the matrix is inserted in fiber form. Polymer fibers are intimately mixed with reinforcement fibers via commingling, wrapping, or co-weaving. The hybrid yarns are then used to form a 2D or 3D hybrid fabric, and the subsequent application of heat and compression melts and spreads the matrix fibers during the consolidation stage. The main advantage of this technique is that, unlike most thermoplastic prepregs, the hybrid yarns and fabrics are highly flexible and drapable, which allows them to perfectly fit a contoured mold with a complex shape. Figure 3.6 illustrates the production of a glass fiber-reinforced polypropylene part starting from a commingled hybrid yarn [109].

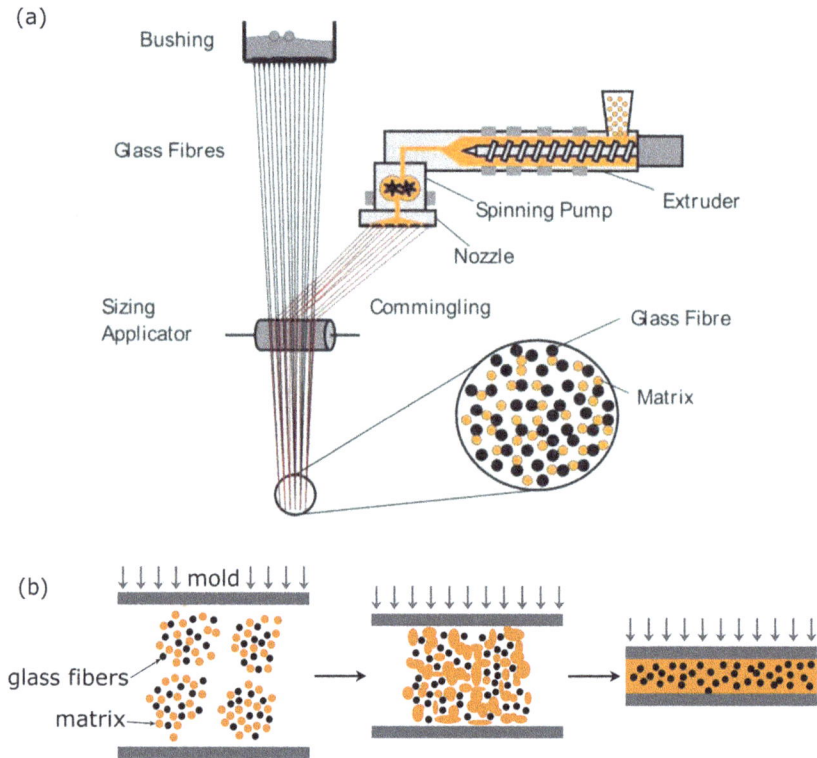

Fig. 3.6: Production and consolidation of commingled hybrid yarns. (a) Example of production of glass fiber/thermoplastic matrix yarns (duplicated from [119]). (b) Compaction process: heat and pressure favor matrix melting and yarn aggregation and bridging, until the matrix flows to impregnate bundles and the subsequent flow on the micro-scale dissolves trapped porosity (adapted from [109]).

The use of hybrid yarns also allows the homogeneous dispersion of other functional fillers in a thermoplastic composite. Conductive particles, sensors, or dyes can be integrated into the matrix filaments or inserted as an additional filament type. This is especially interesting for the development of multifunctional composites, in which the combination of different fillers results in a unique set of properties designed to perform several tasks simultaneously [120, 121].

Finally, liquid impregnation is a technique in which the matrix is supplied as a low-viscosity liquid made of monomers or prepolymers. Due to the low viscosity, it is processable as a thermosetting resin with all the techniques mentioned in Section 3.4.1. Up to some years ago, this process was commonly used only for some thermoplastic polyimides, but the inherent advantages of processing a low-viscosity precursor have paved the way for the development of new formulations (e.g., Arkema's Elium® resin).

3.4.3 Discontinuous-fiber composites

The majority of discontinuous-fiber-reinforced composites have a thermoplastic matrix. The process of making such composites is generally divided into two steps. In the first step, the matrix is compounded with the fibers, generally by means of an extruder, to obtain a uniform dispersion while trying to minimize fiber breakage. In the second step, the compounded material is remelted or softened and molded to obtain the final product. Injection molding is the most common way to obtain short-fiber-reinforced thermoplastic products.

Even though the machines, molds, plungers, and screws can be the same as those used for unreinforced plastics, the processing parameters must be adjusted to deal with the modification of rheological properties and thermal conductivity introduced by the fibers [103]. This is important also because the properties of discontinuous-fiber composites are strongly influenced by the fiber length and orientation, which can both change according to the processing conditions [122].

3.5 Characterization of fiber-reinforced polymer composites

The characterization of fiber-reinforced polymer composites is a critical aspect of materials engineering that enables researchers and industry professionals to understand, predict, and optimize the performance of these advanced materials. Central to the study of these materials is the complex relationship between processing, microstructure, and properties. The manufacturing methods employed, such as hand lay-up, resin transfer molding, or autoclave processing, significantly impact the final microstructure of the composite. This microstructure, in turn, dictates the material's mechanical, thermal, and chemical properties. For instance, the fiber volume fraction, void content, and degree of fiber alignment – all influenced by processing parame- ters – directly affect the composite's strength, stiffness, and durability. Moreover, the processing conditions can influence the quality of the fiber-matrix interface, which plays a crucial role in load transfer and overall composite performance. Understanding and characterizing these interdependencies is essential for optimizing composite materials for specific applications and for developing predictive models of composite behavior.

This subchapter provides an overview of the various techniques and methodologies employed to assess the physical, mechanical, thermal, and chemical properties of these composites. However, providing a comprehensive description of all the techniques and parameters involved in the characterization of these materials is beyond the scope of this book. For more information, the reader is advised to consult other sources, e.g., [123–126].

3.5.1 Microstructural analysis

Microstructural analysis forms the foundation of FRP composite characterization, offering invaluable insights into the material's internal architecture. This subchapter explores the advanced microscopy techniques utilized to visualize and quantify the composite's microstructure.

Scanning electron microscopy (SEM) stands at the forefront of microstructural analysis, providing high-resolution images of fiber orientation, distribution, and interfacial bonding. Through careful sample preparation and imaging protocols, one can observe fracture surfaces, identify failure mechanisms, and assess the quality of fiber-matrix adhesion. Complementary to SEM, transmission electron microscopy (TEM) offers even higher magnification, allowing for the examination of nanoscale features within the composite, such as crystalline structures in the matrix, the distribution of nanofillers in nanoreinforced matrices, or sizing agents on fiber surfaces [127].

Light microscopy, while less sophisticated than electron microscopy, remains a valuable tool for quick assessment of composite cross-sections, especially with polished surfaces. When coupled with image analysis software, optical micrographs can yield quantitative data on fiber diameter, matrix distribution, and interfacial features. Polarized light microscopy offers additional benefits in the examination of optically anisotropic fibers, such as carbon or aramid, by revealing orientation-dependent properties.

X-ray computed tomography (CT) has emerged as a powerful non-destructive technique for the three-dimensional visualization of FRP composites. This method enables the assessment of void content, fiber waviness, and internal defects throughout the bulk of the material. By employing advanced image analysis algorithms, quantitative data on fiber volume fraction, orientation distribution, and pore network characteristics can be extracted from CT scans.

3.5.2 Mechanical characterization

The mechanical properties of FRP composites are of paramount importance in determining their suitability for various applications. This subchapter addresses the wide array of testing methodologies employed to characterize the static and dynamic mechanical behavior of these materials [124].

Tensile testing serves as a fundamental method for assessing the longitudinal strength and stiffness of FRP composites. Through careful specimen preparation and the use of advanced extensometry, accurate stress-strain relationships can be established. Given the anisotropic nature of FRP composites, testing in multiple orientations is required to fully characterize their behavior. The primary standard for tensile testing of polymer matrix composite materials is ASTM D3039/D3039M, which covers the determination of in-plane tensile properties of polymer matrix composite materials reinforced by high-modulus fibers.

Compression testing, while challenging due to the potential for buckling, provides crucial information on the composite's response to compressive loads, particularly important for applications involving structural components. For laminate composites, compression testing can be conducted along the fiber direction, offering valuable insights into the material's behavior under axial compressive loads. This type of testing is particularly relevant for unidirectional composites or for specific plies within a multidirectional laminate. When testing in the fiber direction, special fixtures and specimen designs are often employed to prevent premature failure due to global buckling. The ASTM D3410 standard, for instance, describes a method for compressive properties of polymer matrix composite materials using a shear loading fixture. This approach allows for the determination of key parameters such as compressive strength and modulus in the fiber direction. Another widely used method is the combined loading compression (CLC) test, described in ASTM D6641, which applies both shear and end loading to the specimen, reducing the risk of end-crushing failures. Compression testing along the fiber direction can reveal critical failure modes such as fiber microbuckling, kink band formation, or shear failure of the matrix. These failure mechanisms are often distinct from those observed in tensile testing and provide essential information for designing composite structures that may experience compressive loads during service. Additionally, the compressive strength in the fiber direction is typically lower than the tensile strength, making it a limiting factor in many design scenarios.

Flexural testing, typically conducted in three-point or four-point bending configurations, offers insights into the composite's behavior under combined tensile and compressive stresses. This method is particularly relevant for applications where bending loads are prevalent and is a convenient way to avoid challenges associated with specimen gripping.

Interlaminar shear strength, a critical parameter in assessing the integrity of the fiber-matrix interface, is commonly evaluated through short beam shear (SBS) tests or other specialized methods such as the V-notched beam test.

Impact testing, including methods such as Charpy and Izod tests, as well as drop-weight impact testing, provides essential data on the composite's ability to absorb energy during rapid loading events. These tests are particularly relevant for assessing the material's suitability in applications where impact resistance is critical.

Fatigue testing is crucial for understanding the long-term performance of FRP composites under cyclic loading conditions, which is particularly relevant for applications in the aerospace, automotive, and wind energy sectors. The ASTM D3479 standard, "Standard Test Method for Tension-Tension Fatigue of Polymer Matrix Composite Materials," provides guidelines for conducting fatigue tests on FRP composites. In fatigue testing, specimens are subjected to cyclic loading at various stress levels, and the number of cycles to failure is recorded. This data is typically presented in the form of S-N curves (stress amplitude vs. number of cycles to failure), which provide valuable information about the material's fatigue life and endurance limit. Unlike metallic materials, FRP composites often do not exhibit a clear fatigue limit, and their fatigue behavior can be more complex due to their anisotropic nature and multiple failure modes [128]. The fatigue behavior of FRP composites is influenced by factors such as fiber orientation, matrix properties, fiber-matrix interface strength, and environmental conditions [129]. Common failure modes in fatigue include matrix cracking, delamination, and fiber breakage. These failure modes often interact and evolve over the course of cyclic loading, leading to progressive degradation of the mechanical properties. Advanced monitoring techniques, such as acoustic emission or thermography, can be employed during fatigue testing to detect and track damage evolution. These methods provide insights into the underlying failure mechanisms and can help in developing more accurate life prediction models [123].

3.5.3 Thermal and thermomechanical characterization

The thermal and thermomechanical properties of FRP composites play a crucial role in characterizing the properties of the matrix like the degree of crystallinity or the quality of the curing process, as well as in determining the performance of the composites in high-temperature environments and under thermal cycling conditions. This subchapter explores the techniques used to evaluate these properties and their implications for material behavior.

Differential scanning calorimetry (DSC) stands as a primary tool for investigating the thermal transitions within the polymer matrix of FRP composites. By measuring heat flow as a function of temperature, DSC can reveal important parameters such as glass transition temperature (T_g), melting temperature, and degree of cure of the polymer matrices. These properties have significant implications for the composite's mechanical behavior and long-term stability [124].

Thermogravimetric analysis (TGA) complements DSC by providing information on the material's thermal stability and decomposition behavior. TGA data is invaluable for assessing the upper temperature limits of the composite and for studying the kinetics of thermal degradation. When coupled with mass spectrometry or Fourier-transform infrared spectroscopy, TGA can offer insights into the nature of volatile products evolved during thermal decomposition. When the reinforcing fibers do not degrade in the considered temperature interval, the residual mass after the test can be used to calculate the experimental fiber weight fraction.

Coefficient of thermal expansion (CTE) measurements are critical for understanding how FRP composites respond to temperature changes, particularly in applications where dimensional stability is crucial. Techniques such as thermomechanical analysis (TMA) or dilatometry allow for precise measurement of CTE in different material directions, highlighting the anisotropic nature of thermal expansion in these composites.

Finally, dynamic mechanical thermal analysis (DMTA) has proven invaluable in characterizing the viscoelastic properties of FRP composites. By subjecting specimens to small oscillatory loads over a range of temperatures and frequencies, researchers can probe the material's storage modulus, loss modulus, and damping characteristics. This technique is particularly useful for identifying secondary transitions in the polymer matrix and for studying the temperature dependence of mechanical properties. DMTA data can also be used to construct master curves, enabling the prediction of long-term mechanical behavior through time-temperature superposition principles.

3.5.4 Chemical and environmental characterization

The long-term durability and performance of FRP composites in various environments are critical considerations for many applications. This subchapter addresses the methods used to assess the chemical stability and environmental resistance of these materials.

Fourier-transform infrared spectroscopy (FTIR) serves as a powerful tool for analyzing the chemical composition and structure of both the matrix and fiber components in FRP composites. FTIR can be used to monitor changes in chemical bonding resulting from environmental exposure, providing insights into degradation mechanisms. Attenuated total reflectance (ATR) FTIR offers the additional advantage of surface-specific analysis without extensive sample preparation.

X-ray photoelectron spectroscopy (XPS) offers detailed information on the surface chemistry of FRP composites, including the nature of fiber-matrix interactions and the presence of contaminants or treatment agents. This technique is particularly valuable for studying the effects of surface treatments on fiber-matrix adhesion and for investigating environmental degradation mechanisms.

Environmental aging studies form a crucial component of FRP composite characterization. These typically involve exposing materials to controlled environments simulating service conditions, such as elevated temperatures, humidity, UV radiation, or chemical agents. Subsequent analysis of mechanical properties, chemical composition, and microstructure allows for the assessment of degradation mechanisms and the prediction of long-term performance.

Moisture absorption kinetics and their effect on composite properties are often studied through gravimetric methods coupled with mechanical testing. Understanding the diffusion behavior of water in FRP composites is essential for predicting their performance in humid environments and for designing appropriate protective measures [127].

4 Multifunctional composites and structural TES materials

The previous chapters have introduced the concept of thermal energy storage and the principal materials and technologies through which it is translated into real applications. An introduction to polymer composites was also presented, illustrating the main polymer matrices, reinforcement phases, and production processes, as well as their potential for lightweight design.

As the general aim of this book is to present the possibility of integrating TES technology into a structural material to produce a multifunctional component, the present chapter introduces the concept of multifunctional material and explains the role of polymer composites in the development of multifunctionality.

4.1 Multifunctional materials

A material can be classified as multifunctional if it features a set of properties that make it suitable to simultaneously sustain stimuli of different natures and to be ready to perform different functions when required. Multifunctional materials can combine a set of attributes chosen among the structural mechanical properties, like stiffness, strength, and toughness, while non-structural features encompass sensing and actuating capabilities, optical and magnetic qualities, self-repair mechanisms, thermal and electrical conduction or insulation, resistance to corrosion, friction-related properties, and the ability to harvest and store energy [130, 131].

These innovative materials have significant potential to revolutionize future structural performance by decreasing weight, volume, expenses, and energy usage while boosting efficiency, adaptability, and safety. As they broaden design options and increase the value of materials, they are garnering growing attention from both industry and academia. The next generation of multifunctional materials incorporates novel property combinations, making them suitable for use in the automotive and aerospace sectors, as well as in civil engineering and medical applications [130].

The combination of properties of a multifunctional material should act synergistically and not parasitically. If the addition of self-sensing and self-healing capabilities to a structural material impairs its stiffness and strength excessively, the combination of properties in this material will not bring benefits at the system level. To achieve this, multifunctionality should be considered from the very early stages of the material design process. Unlike natural materials, whose properties are the result of a locally random evolution process, in synthetic engineering materials the design must start from the functions and proceed in a specific direction. This is important because multifunctionality could be expressed at different length scales (at the phase, mate-

https://doi.org/10.1515/9783111111865-004

rial, or structure level), which must be taken into account during the design of the material and the prediction of the final properties of the component [132].

4.1.1 Multifunctionality of polymer composites

Multifunctional design is particularly well-suited for composite materials, as they can incorporate the characteristics of multiple components into a single material. Numerous natural and synthetic substances that need to fulfill multiple roles are composed of polymer-matrix composites. These materials effectively combine exceptional polymer-related qualities, such as durability and lightweight properties, with the specific attributes of the discontinuous phase(s). This combination not only provides stiffness and strength but also offers various non-structural functionalities.

Multifunctional composites can perform multiple structural functions or combined structural and non-structural functions. Most of the recent developments in multifunctional materials tend to be polymer-matrix structural composites featuring one or more additional non-structural functions. This strategy allows large weight savings at the system level through the elimination or reduction in the number of multiple monofunctional constituents [131], and it gives better results than the conventional approach of optimizing the weight and geometry of the single subsystems individually.

Composite materials can also reach a high level of multifunctionality by combining reinforcements on different scales, which range from continuous fibers to particles at the micro- and nanoscale [130].

4.1.2 Applications and examples

As described in Chapter 3, polymer composites have become the materials of election for lightweight structures. As the employment of multifunctional composites can lead to considerable weight savings, these materials have become attractive because, besides their remarkable specific stiffness and strength, they open the possibility of adding other functionalities with a negligible weight and volume increase.

One of the most interesting examples of the combination of structural and non-structural functions is provided by structural batteries, i.e., devices that can carry mechanical load and store electrical energy at the same time [131]. The concept of structural batteries is a groundbreaking design approach that integrates energy storage directly into a component's structure, effectively using a battery to construct structural elements. This integration results in a total device mass that is less than the combined weight of separate structural components and batteries. For example, a mass saving of up to 20% can be achieved in an electric car by building the roof panel with a structural battery instead of having a traditional roof panel and a traditional lithium-ion battery [133]. As the world increasingly shifts towards electrification, this innovative approach is particularly beneficial for applications requiring portability and power, such as electric vehicles and handheld or wearable electronics. The effectiveness of this concept increases as the multifunctional composite's structural and energy storage capabilities approach those of conventional structural materials and traditional batteries, respectively. However, in practice, structural batteries typically do not match the mechanical performance of purely structural components or the electrochemical properties of standard batteries. Nevertheless, the advantages of multifunctionality and the overall mass and volume reduction must be evaluated at the system level to fully appreciate their benefits.

Several approaches have been taken to develop structural batteries. At the forefront of this field are batteries based on carbon fiber-based electrodes (CFBEs) (Fig. 4.1), which offer promising improvements in both electrochemical and mechanical performance compared to traditional battery components. In this design concept, carbon fibers are part of both the cathode and the anode.

Being able to intercalate lithium ions in their microstructure, carbon fibers can act as an anode material. Thanks to a combination of high reversible capacity and high stiffness and strength, which results from their microstructure rich in turbostratic disordered carbon, intermediate modulus polyacrylonitrile (PAN)-based carbon fibers are the most widely used anode materials in structural batteries [134]. Their capacity can be further increased by coating them with high-capacity conversion-type materials like metal oxides: in this way, reversible capacities over 600 mAh/g have been achieved – significantly higher than the theoretical capacity of graphite anodes used in conventional lithium-ion batteries. The carbon fiber substrate helps mitigate the volume expansion and pulverization issues that normally plague conversion-type anodes. Carbon fiber anodes have also shown good rate capability, with some electrodes maintaining high capacities even at fast charge/discharge rates above 3C. However, the coating process can negatively impact the mechanical properties of the carbon fibers. While some coatings like reduced graphene oxide have been found to improve tensile strength, many decrease both strength and stiffness compared to pristine fibers. Balancing the electrochemical and mechanical performance remains an ongoing challenge, with factors like coating thickness, uniformity, and choice of materials all playing important roles.

In carbon fiber-based cathodes (CFBCs), carbon fibers are primarily used as current collectors in cathodes, where they support intercalation-type materials like lithium-iron-phosphate (LFP) and nickel-manganese-cobalt (NMC). These materials are coated onto carbon fibers to enhance their electrochemical performance, although the capacity of CFBCs is generally limited compared to anodes. However, mechanical property studies of CFBCs are limited, with only adhesion between the coating and carbon fiber examined so far. More research is needed to optimize the balance between electrochemical and mechanical performance for these structural cathode materials.

While most CFBE research has focused on lithium-ion technology, other battery chemistries have also been explored to bring the benefits of structural power to different systems. Zinc-manganese dioxide structural batteries have demonstrated promising electrochemical and mechanical properties. Sodium-ion batteries using carbon fiber electrodes have shown improved performance compared to conventional electrodes. Carbon fibers have also been investigated as flexible electrodes for lithium metal batteries and as strain sensors through the piezo-electrochemical transducer effect. These alternative chemistries expand the potential applications for structural power composites beyond traditional lithium-ion systems.

Producing full cells with CFBEs presents several challenges that require further research. The interface between the electrode coating and the surrounding solid electrolyte matrix is not well understood and needs investigation to optimize both mechanical and electrochemical performance. Current CFBE characterization methods using chopped fibers in coin cells are not representative of real structural battery architectures. Future studies should test continuous fiber CFBEs in biphasic electrolytes to better reflect actual cell designs. Full cell assembly would also help address issues like electrode balancing and volume changes during cycling. Additionally, recycling CFBEs is particularly challenging and requires developing methods to recover carbon fibers without compromising their properties.

In conclusion, carbon fiber-based electrodes show significant promise for increasing the energy density of structural batteries beyond current levels. However, more research is needed on mechanical properties during cycling and optimizing the electrochemical-mechanical performance balance. Assembly and recycling challenges also need to be addressed. Overall, CFBEs offer a pathway to create high-performance structural power composites for demanding applications like electric aircraft, but further development is required to realize their full potential. Future work should focus on sustainable materials, modular designs for easier recycling, and novel coating techniques to improve the overall sustainability and cost-effectiveness of these multifunctional composites.

When designing a structural battery like this, challenges arise as the optimization of the mechanical properties of a phase often causes a decrease in its electrochemical performance. For example, a solid polymer electrolyte exhibiting high ion conductivity often shows poor stiffness and strength, and highly graphitized carbon fibers featuring excellent stiffness are not the optimal choice from the point of view of the Li-ion intercalation efficiency [134]. It is therefore important to fully understand how the microstructure of each phase correlates with the mechanical and electrochemical properties, as each component should be tailored to reach the best property set, and this must be taken into account in the early stages of the fabrication of each phase.

Another class of materials generally classified as multifunctional is that of the "smart" materials, which are able to adapt and respond to external stimuli by performing both as sensors and as actuators [135]. The state-of-the-art smart materials can change their properties (e.g., mechanical, electrical, magnetic, rheological), color,

(a)   (b)

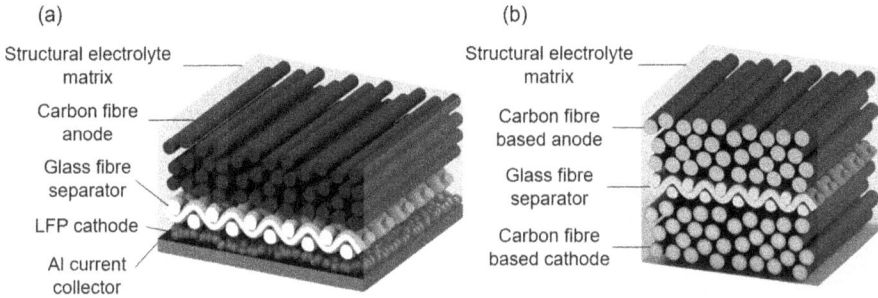

Fig. 4.1: Current state-of-the-art structural battery and an innovative structural battery concept with a laminated architecture using carbon-based electrodes (duplicated with permission from [133]).

or shape depending on external stimuli such as heat, electricity, light, solvent, and pH value. Several applications of such materials can be found in buildings and automotive/aerospace fields, to actively control vibrations, noise, and deformations [135].

A powerful example of this concept is represented by multifunctional concrete, which is increasingly attractive for civil engineering applications as it integrates traditional concrete's structural capabilities with additional functionalities, such as electrical conductivity and sensing properties. This innovative composite material is particularly useful in various applications, including structural health monitoring, energy conversion, and environmental protection.

For example, self-sensing concrete exhibits a sensible change in electrical resistivity with applied stress or strain, which makes it useful for structural health monitoring [136]. One of the most frequently pursued strategies to produce self-sensing concrete is to embed conductive fillers (e.g., CNTs and xGnPs) at or near the percolation threshold. In this way, the formation of damages and cracks interrupt the conductive paths, thereby causing a remarkable increase in electrical resistivity [137]. Moreover, some conductive additives such as the xGnPs have proven to enhance the durability and the compression and flexural strength of the host concrete, thus being a multifunctional filler themselves [137]. Advanced techniques, such as hierarchical structuring, further enhance this self-sensing ability by optimizing the distribution and connectivity of the conductive fillers, ensuring that even minor changes in stress can be detected effectively.

Despite its potential, the design and implementation of multifunctional concrete face several challenges. Achieving a balance between the mechanical properties and the desired electrical characteristics can be difficult, as increasing the amount of conductive fillers may compromise the concrete's structural integrity. Additionally, environmental factors can affect the performance of self-sensing capabilities, and the lack of established regulatory standards for these materials can hinder their widespread adoption in construction projects. Furthermore, the high costs associated with developing and maintaining these advanced materials pose significant barriers to their practical application [138].

In conclusion, multifunctional concrete represents a significant advancement in construction materials, offering numerous benefits for infrastructure development. As research continues to address the challenges associated with its design and implementation, there is strong potential for broader applications in smart cities and sustainable construction practices. Future perspectives include enhancing the understanding of the underlying mechanisms of multifunctional properties and developing standardized guidelines for its use, which will ultimately facilitate its integration into large-scale engineering projects [138].

The same self-sensing concept can be applied to polymer composites. Several examples have been proposed of continuous-fiber composites containing CNTs, either dispersed in the matrix or deposited onto the fiber surface. In such composites, the electrical conductivity is dominated by the formation of a percolative path of CNTs, and this is the case also when the reinforcement is made of carbon fibers, due to the remarkably higher electrical conductivity of CNTs compared to CFs [139, 140].

4.2 Structural TES polymer composites

The previous sections showed that composites in general, and polymer composites in particular, exhibit great potential to be designed as multifunctional, as their properties are the result of the combination of two or more phases. This can further expand the weight-saving possibilities pursued when composites are selected as structural materials. The scientific literature contains many examples of multifunctional composites combining structural properties and a wide range of non-structural functions, which respond to a wide variety of needs in the most diverse applications.

An interesting field in which the multifunctionality of polymer composites can be exploited is that of structural composites with thermal energy storage capability. As explained in the previous sections, among the applications of TES materials are (i) heat storage for temperature control, for example in the building industry, or to produce smart textiles for body temperature regulation, and (ii) the temporary storage of heat to prevent overheating, as in cooling systems for electronic devices [5].

When a TES material is used for thermal management, it is usually only an extra component added to the main structure of a device. For example, Liu et al. [141] developed a novel refrigeration system for refrigerated trucks that incorporates a low-cost PCM to maintain desired thermal conditions. The employed PCM had a melting temperature of approximately −26.7 °C and a latent heat of fusion of 154.4 J/g, making it suitable for maintaining the desired temperature of −18 °C in the refrigerated space. It was encapsulated in plastic pouches and incorporated into a phase change thermal storage unit (PCTSU), located outside the refrigerated chamber. The system utilizes an off-vehicle refrigeration unit to charge the PCM, which then discharges cooling energy during transport, by absorbing latent heat while melting and keepint the indoor temperature low. The findings indicated that this innovative system can achieve energy cost savings of up to 86.4% compared to conventional systems, while also producing lower greenhouse gas emissions and improved temperature control. The system is designed to be versatile and suitable for both small and large refrigerated trucks. The amount of PCM used is sufficient to maintain the desired temperature for up to 10 h, even in the hottest climatic conditions. However, the study also highlighted potential disadvantages, such as the need for added PCM-related weight and enhanced insulation to maintain the refrigerated space at −18 °C for extended periods. Additionally, safety considerations regarding the PCM and heat transfer fluid were addressed, emphasizing the importance of proper maintenance to ensure safe food transportation. The drawbacks highlighted by Liu et al. in using PCM-based refrigerated systems – additional weight, safety issues, short distance only – are generally recognized as the three main issues holding back the wide utilization of this refrigerating system in transportation [142].

In particular, the resulting increase in weight and volume may be unacceptable for some applications, such as refrigerated transportation, where weight and volume reduction are crucial design parameters. In this case, it would be useful to have a multifunctional material combining good mechanical properties and the heat storage/management function. With an approach similar to that developed with structural batteries, such materials could be used to build part of the structure with the "thermal battery" material, or, in other words, to design a structure that is part of the thermal management system. According to the author's opinion, these structural TES lightweight composites can be attractive in four main applications [5]:

1. In the field of transportation, polymer composites are increasingly employed to build structural or semi-structural components, thanks to their high specific mechanical properties. However, the considerable application of polymer composites in replacement of traditional materials could complicate thermal management in the indoor vehicle environment, due to the different thermal conductivity and heat capacity. This could result in increasing difficulty in maintaining the indoor temperature within the human thermal comfort range. This problem is further exacerbated by the diffusion of electric vehicles, in which indoor thermal comfort cannot even be based on the excess heat produced by the thermal motor. This issue could be addressed by introducing a TES system able to store and release thermal energy in the human comfort temperature range, and the overall performance-to-weight ratio would be maximized if this TES material is part of the structural components themselves. This concept could be extended to other temperature ranges, which may be interesting for the refrigerated transportation of food, biomedical items, or other perishable goods.

2. In the field of portable electronics, there is an increasing tendency to enhance the performance, power, and functionalities of the devices while reducing weight and volume. This trend brings issues in the thermal regulation of the device, especially during peak power operations. As described previously, passive cooling systems based on

PCMs are attracting attention due to their capacity to maintain the temperature in the desired range – more specifically, under a certain threshold value – and to cope with the momentary but rapid heat generation during peak power operations. However, in all the solutions implemented so far, the PCM is contained in an additional module, while it would be advantageous to embed it directly inside the structural components such as phone or laptop cases, or electric vehicle battery cases. Another interesting application in the electronics field would be the production of PCM-enhanced circuit boards, traditionally made of glass fiber-reinforced epoxy resin.

3. Among the main drawbacks of polymer-matrix composites, especially when compared to metals or ceramics, are the lower thermal resistance and the loss of mechanical properties with increasing temperature. This can be detrimental for some applications where the composite material is subjected to external heating in service, as in the case of luxury car chassis entirely made of carbon fiber composite. However, local overheating and loss of properties could also be determined by dynamic loading and lack of heat dissipation, which can lead to premature failure by fatigue. This effect, critical especially for thermoplastic composites, could be limited by adding a PCM, which absorbs excess heat and prevents temperature rise. In this way, if the heat cannot be dissipated, it may at least be absorbed, thereby preventing a temperature rise. This concept has been proven effective by Casado et al. [143], who inserted a hydrated salt with a transition temperature of 50 °C in a polyamide-based composite reinforced with glass fibers. In this case, however, the PCM was not employed as an additional filler, but it was added in the gaps of the flanged plate made of the glass-reinforced polyamide. Much more capillary would be the action of the PCM if it were dispersed in the whole composite mass, or at least in the parts more subjected to fatigue damage.

4. One of the main problems of outdoor infrastructures operating in cold environments is related to the formation and accretion of ice. Critical structures such as transmission lines, wind-driven power generators, off-shore oil rigs, and means of transportation like aircraft and ships can be damaged by the excessive weight of the ice layer and the stress induced by the freeze-thaw cycles [144]. The solutions implemented so far rely on heating methods, mechanical methods, or the circulation of de-icing fluids underneath the surface, which are effective but extremely energy-consuming [145]. More recently, slippery or superhydrophobic coatings have been developed that prevent ice adhesion or reduce ice shear strength, but they generally have low resistance to mechanical abrasion and poor durability to outdoor weather agents. Alternatively, for the polymer composite structures, e.g., the wind turbine blades, an effective alternative could be the introduction of a PCM with phase transition within the range of −10 °C to 0 °C in the whole composite thickness or only in the outermost layers [146]. The ability to maintain the temperature stable during phase transition can effectively delay the freezing of water droplets on surfaces, which is essential in preventing ice formation. Additionally, if the PCM is added in its "free" (non-microencapsulated) form, the smoothness of the liquid film created by PCMs can increase the energy barrier for ice nucleation, further

reducing the likelihood of ice accumulation. Recent advancements have also explored the integration of PCMs with photothermal materials, enhancing the melting of ice through rapid heating while managing temperature effectively [147].

In all these applications, PCMs are the ideal TES materials, as they work at a nearly constant temperature and exhibit a high enthalpy per unit mass, thus being suitable for applications where weight saving is a key factor. It is therefore fundamental to understand how the PCM addition influences the mechanical properties of the host composite and how this effect varies below and above the PCM phase change.

So far, this investigation has been carried out mostly on construction materials like concrete and gypsum, in which a PCM phase is added to enhance the thermal management of indoor environments while reducing the energy consumption for indoor heating/cooling. As described in Section 2.7.1, the majority of TES systems integrated into walls or flooring materials are represented by organic PCMs with a phase transition temperature in the range of 18–25 °C, microencapsulated in polymeric shells. The characterization of such PCM-enhanced construction materials generally evidences that an increase in the PCM fraction brings an increase in the thermal management properties and a decrease in the heating/cooling power consumption. On the other hand, PCM addition determines a decrease in compressive and flexural properties, due to the introduction of a less mechanically strong phase. Therefore, the authors of the revised papers generally conclude that it is important to select the PCM causing the smallest decrease in mechanical properties while providing the highest TES capability, and that it is fundamental to identify the optimal PCM content to obtain a material with the most suitable property set for a specific application [148].

The open scientific literature provides a good number of examples of polymers containing a PCM, but fewer cases of PCM-enhanced polymers employable as structural materials. Subchapter 4.3 reports some examples of PCM-containing polymers and polymer composites, with the main investigation techniques and the principal conclusions.

4.3 PCMs in polymers and polymer composites

The present subchapter reports examples of variously stabilized PCMs contained in polymer matrices and polymer composites. The research was narrowed to the PCMs with a phase change temperature in the low-medium temperature range (15–80 °C), as this is the interval of interest for the aforementioned applications of structural TES composites. In this temperature range, organic PCMs dominate the market and academic research, and thus the present chapter focuses on this PCM class and specifically describes how the added PCM affects the host polymer and composite, by presenting the general principles and some case studies.

4.3.1 PCMs in polymer matrices

4.3.1.1 Blends between PCMs and polymer matrices

The scientific literature reports many examples of polymer matrices containing organic PCMs (mostly paraffins) without further stabilization. In this way, a sort of polymer/ PCM blend is obtained. This is commonly referred to as the simplest and cheapest way to avoid leakage of the PCM while defining the shape of the PCM-containing component. Due to the absence of leakage and of traditional macro- or micro-encapsulation, these blends are sometimes referred to as "solid-solid PCMs." However, unlike the "real" solid-solid PCMs (or solid-state PCMs), which rely on a solid-solid phase transition [68] or in which liquefaction is prevented by grafting or crosslinking [149], in these polymer/PCM blends the PCM still undergoes the same melting transition (a solid-liquid phase change), but no liquid is visible from the outside because it is confined by the surrounding polymer matrix. Such polymer/PCM blends are also sometimes called "matrix-encapsulated PCMs," again to refer to the fact that the surrounding polymer matrix prevents leakage of the PCM above the melting point. To avoid confusing terms, and following the definitions given in the first chapters of this book, the terms "matrix-encapsulated PCMs" and "solid-solid PCMs" will not be used here to refer to PCM/polymer blends, which will be defined here simply as a particular form of shape-stabilized PCMs.

A large number of polymers have been used as supporting materials, such as traditional thermoplastics (PE, PP, polyamide (PA), polymethylmethacrylate (PMMA)), thermosets (epoxy) and elastomers (styrene block copolymers, ethylene-propylene-diene-monomer (EPDM)).

Chen and Wolcott [150, 151] blended a paraffinic PCM with high-density, low-density, and linear low-density polyethylenes (HDPE, LDPE, and LLDPE) via melt mixing, and the total paraffin content was fixed at 70 wt% in all three cases. Morphological investigations showed the formation of a co-continuous structure between the two phases; this was considered the main cause for the paraffin leakage from the prepared blends, which was up to 10 wt% of the total initial paraffin fraction. The phase change enthalpy and temperature of the paraffin varied only slightly after blending, while the PE crystallinity decreased sensibly, especially for LDPE and LLDPE. The authors focused on the miscibility of the PCM with the polymer matrices and the blend morphology, but they did not investigate the mechanical properties nor the long-term stability of the prepared samples. This is a crucial aspect even if the materials are not meant to be applied under heavy load-bearing conditions, as underlined by other authors [55, 152] who investigated the thermo-physical and mechanical properties of polyolefins/paraffin blends. These blends contained a paraffin weight fraction of up to 38 wt% and a melting enthalpy of up to 71 J/g. A slight alteration in the matrix crystallinity upon blending with paraffin was detected also by these authors, who attributed the decreased melting temperature of the host polymer to a plasticizing effect of the PCM, especially for the samples without co-crystallization (PCM/polypropylene (PP)). The authors highlighted that the mechanical performance of the host matrices, measured under quasi-static and dynamic conditions, decreased with an increase in the paraffin concentration, and this was even more evident above the PCM melting temperature. The authors concluded that such blends cannot be applied as load-bring components but only as functional materials.

To increase the mechanical properties of the polymer/PCM blends, some authors proposed the addition of fillers such as EG, xGnP, or nanoclays, which are also helpful in enhancing thermal conductivity. This was proved by Sobolciak et al. [153, 154], who investigated the thermomechanical performance of paraffin/LLDPE blends containing EG microparticles and exhibiting a phase change enthalpy of up to 31 J/g. For the blends containing 15 wt% of EG, the storage modulus was doubled compared to that of the neat LLDPE/paraffin blends, and the thermal conductivity increased from 0.25 to 1.32 W/(m · K). The same concept was applied by Wu et al. [155], who designed an EG/paraffin/olefin-block-copolymer (OBC) composite with thermally induced flexibility. The authors conclude that the addition of EG not only contributes to the mechanical stability of the composite but also helps in preventing leakage due to the capillary force and surface tension. Similar observations were made by Fredi et al [156], who incorporated a micrometric powder made of a paraffin wax shape-stabilized with CNTs in an epoxy resin. They first optimized the CNT content to 10 wt% to effectively prevent paraffin leakage above its melting point, then dispersed this CNT-stabilized paraffin powder into epoxy resin at various concentrations (20–40 wt%). It was found that the CNT-stabilized blends maintained 80–90% of their theoretical melting/crystallization enthalpy and showed stable performance over 50 thermal cycles, while blends without CNTs suffered significant paraffin exudation during processing. The addition of the CNT-stabilized paraffin decreased the epoxy's flexural modulus and strength according to the rule of mixtures, though interestingly, the strain at break showed a positive deviation from theoretical predictions. The presence of CNTs also imparted electrical conductivity to the blends, with resistivity reaching $1.2 \times 10^3 \ \Omega \cdot \mathrm{cm}$ at 40 wt% paraffin + CNT content.

4.3.1.2 Microencapsulated PCMs in polymer matrices

PCMs can also be added to polymer matrices in the form of microcapsules and, in this case, they are added like any other fillers in the production process. The capsule shells provide unmatched PCM containment and are more effective in avoiding leakage compared to any other shape-stabilization techniques. However, this holds only if the processing conditions of the host polymer matrix are mild enough to preserve the shell integrity, as high temperatures, pressures, and shear stresses can cause cracks and damage in the thin capsule shell.

This explains why it is particularly challenging to introduce PCM microcapsules in traditional thermoplastic matrices. For example, Zhang et al. [157] produced core-sheath polyethylene fibers containing microencapsulated n-octadecane via a melt-spinning process, for textile applications. The capsules, having a diameter of 0.4–

4.5 μm, were confined in the fiber core, while the sheath was entirely made of PE and accounted for 60% of the total fiber mass. The produced filaments had a melting enthalpy of up to 13 J/g and mechanical properties suitable for fabricating textile products. The authors highlighted that the processing parameters during melt blending and melt spinning were fundamental for determining the capsule integrity and thus the final total phase change enthalpy.

More recently, Krupa et al. [158] inserted paraffin microcapsules in an HDPE matrix. The authors synthesized paraffin microcapsules with a 1.5-μm thick melamine-formaldehyde shell and a total melting enthalpy of approximately 60 J/g. The capsule introduction caused a considerable decrease in all the mechanical properties of HDPE (elastic modulus, stress and strain at yield and break strain at yield and break). Although the shell material is stiffer than HDPE, its small thickness caused the capsules to deform when the composite was strained, which, together with the insufficient interfacial adhesion to HDPE, prevented the capsules from acting as a reinforcement. Also in this case, the authors concluded that the produced blends retained sufficient mechanical properties to be safely manipulated and applied in real service conditions, but they were not suitable to be employed as load-bearing components.

Thermosetting materials such as epoxies have also been enriched with microencapsulated or variously stabilized PCMs. The advantage of these materials compared to thermoplastics is that the processing starts from low-viscosity liquids and is generally performed at room temperature. This helps to preserve the capsule integrity and avoid PCM degradation, thereby enhancing the phase change enthalpy at the end of the process. An important contribution to the study of the thermo-mechanical behavior of epoxies containing paraffin microcapsules was provided by Su et al. [159, 160], who focused on the microcapsule/epoxy interfacial adhesion. By subjecting their epoxy/microencapsulated PCM composite to repeated thermal cycles, the authors observed that the expansion and shrinkage of the microcapsules and the surrounding matrix occurred with different thermal expansion coefficients. This is common in materials composed of several different phases, but with PCMs the situation is complicated by the additional volume variation caused by the phase change. Even though the volume variation is not as enormous as in the case of a liquid-to-vapor phase change, it must be considered since it can undermine the interfacial stability of the composite and thus its overall mechanical performance. As represented in Fig. 4.2, these phenomena may cause microcracks and fractures in the matrix, which can not only lead to PCM leakage but also to a decrease in the mechanical properties of the overall material.

The same authors synthesized melamine-formaldehyde-based microcapsules and observed that the addition of methanol during the synthesis led to higher interface bonding and enhanced ductility of the interphase region. They also concluded that the stability behavior was greatly influenced by the thermal cycling conditions, as higher heating rates caused higher interfacial damage. The authors formulated a theoretical approach to model the mechanical strength of the composite as a function of

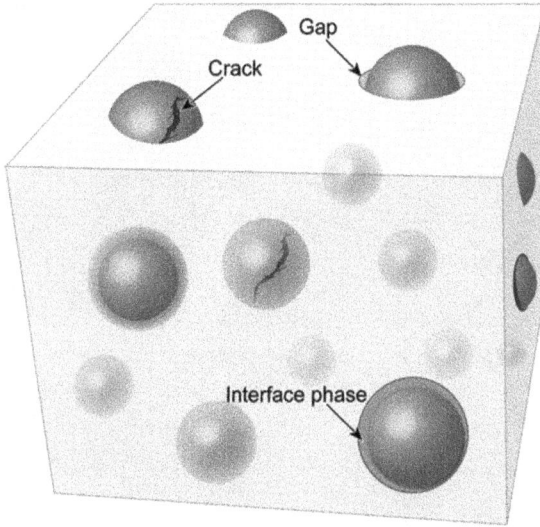

Fig. 4.2: PCM microcapsules in a polymer matrix (duplicated with permission from [159]).

the thickness and integrity of the capsule-matrix interphase region, but they recognize that it is generally difficult to model the mechanical properties of such composites in normal service conditions, especially when the thickness of the interphase region is not clearly measurable [161].

Other authors have tried to increase the interfacial bonding between PCM microcapsules and the surrounding epoxy matrix. For example, Peng et al. [162] enhanced the interfacial bonding between the PCM microcapsules and the epoxy matrix by doping modified TiO_2 nanoparticles into melamine formaldehyde (MF) shell materials. They first modified TiO_2 nanoparticles with 3-Glycidoxypropyltrimethoxysilane (KH560) to introduce epoxide groups on the nanoparticle surface, then incorporated these modified nanoparticles into the MF prepolymer during in-situ polymerization to create a hybrid shell encapsulating n-octadecane. SEM analysis of composite fracture surfaces (Fig. 4.3) revealed that unmodified microcapsules cleanly debonded from the matrix, while the hybrid shell microcapsules showed shell fragments remaining in the matrix – indicating a shift from interfacial failure to cohesive failure of the microcapsule shell. This improved adhesion was attributed to covalent bonding between the epoxy groups on the modified TiO_2 particles and the epoxy matrix through the amino groups of the curing agent. The enhanced interfacial bonding, combined with a 30.4% increase in microcapsule rupture strength, resulted in composites with 17.2% higher tensile strength at 10 wt% microcapsule loading compared to composites containing unmodified microcapsules.

In recent years, many authors have explored different strategies to modify the surface of PCM microcapsules to increase the interfacial adhesion with a polymer (espe-

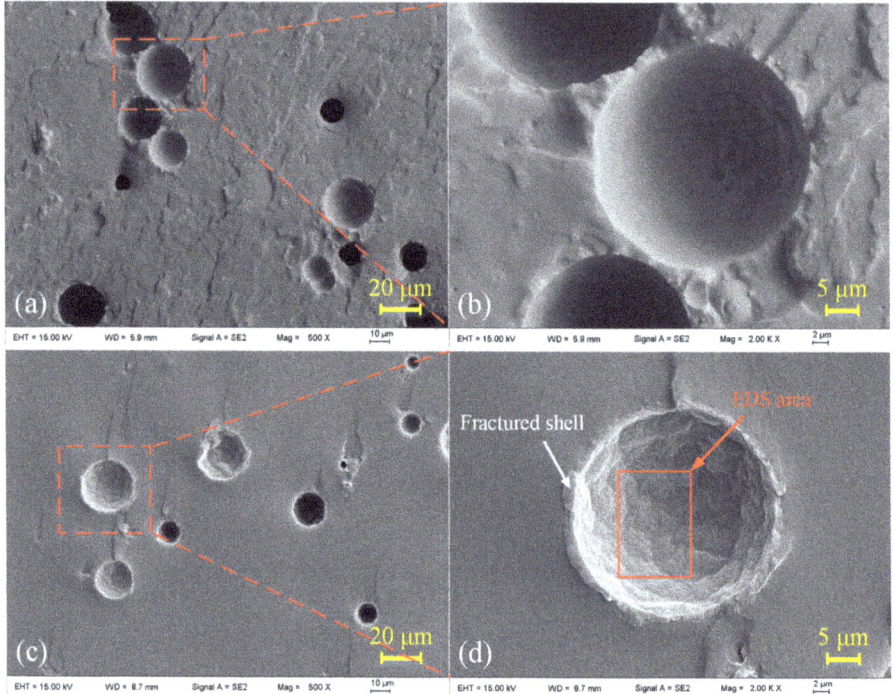

Fig. 4.3: SEM micrographs of (a–b) the tensile fracture surface of the non-modified MF PCM/epoxy composites, with evident PCM debonding, and (c–d) the tensile fracture surface of the TiO$_2$-modified PCM/epoxy composites with evidence of broken PCM microcapsules (duplicated with permission from [162]).

cially, epoxy) matrix. PCM microcapsules with an MF shell are the most common on the market, but a fully cured melamine-formaldehyde surface is not very easy to modify *a posteriori*, especially with traditional surface modification techniques like silane chemistry, because of the lack of reactive functional groups. Hence, strategies have been adopted to modify the MF shell during the encapsulation process itself, similarly to the previously presented approach described in the paper by Peng et al. [162].

Among the possible treatments that have been applied to MF-shelled PCM microcapsules *a posteriori*, a method that is particularly interesting and easy to apply is the deposition of polydopamine (PDA), a synthetic polymer inspired by biological systems. Formed through the self-polymerization of dopamine, PDA contains catechol and amine groups, resulting in a chemical composition similar to that of mussel adhesive proteins. PDA enables the creation of continuous layers with relative ease and remarkable adaptability, demonstrating excellent adhesion to a wide range of metallic, polymeric, and ceramic surfaces. PDA is particularly useful for functionalizing chemically inert surfaces and has been utilized to enhance the fiber/matrix interfacial adhesion of highly inert reinforcing fibers, such as ultra-high-molecular-weight polyethylene (UHMWPE) and poly(p-phenylene-2,6-benzobisoxazole) (PBO) fibers. This bioinspired coating's unique chemistry and diverse array of reactive functional groups allow it to serve as a compatibilizer in various filler/matrix systems and undergo subsequent functionalization. PDA has recently been used also to functionalize PCM microcapsules. For example, Fredi et al. [163] investigated the use of PDA coating to enhance the adhesion between commercial paraffin-based PCM microcapsules and an epoxy matrix (Fig. 4.4a–b). They found that the PDA coating significantly increased surface roughness from 9 to 86 nm and introduced reactive functional groups capable of forming covalent bonds with the epoxy matrix's oxirane groups. Importantly, DSC and hot-stage optical microscopy confirmed that the coating process did not compromise the PCM's phase change properties. SEM analysis of cryofractured composite samples revealed markedly improved interfacial adhesion between the PDA-coated microcapsules and the epoxy matrix compared to uncoated ones, with the coating thickness measured at 265 ± 85 nm. The researchers attributed this enhancement to both mechanical interlocking from the increased surface roughness and chemical bonding between the PDA layer and the matrix, demonstrating that PDA coating is an effective method for improving the typically poor adhesion between formaldehyde-based PCM microcapsule shells and epoxy matrices. A similar PDA-modification approach was used to prepare flexible PCM/epoxy composites with enhanced mechanical properties compared to unmodified PCM/epoxy samples [164] (Fig. 4.4c). Also in this case, the property enhancement was attributed to an increased surface adhesion between the PCM and the surrounding matrix. In this work, by tuning the PDA deposition parameters, a much thinner PDA layer was added (53 ± 8 nm, total PDA mass 2.2 wt %), which proved equally effective in enhancing the interfacial adhesion and limited the decrease in melting enthalpy of the PCMs. In fact, the phase change enthalpy of MC-PDA was only slightly lower than that of the neat MC, being 221.1 J/g and 227.7 J/g, respectively.

PDA functionalization was also employed as an intermediate layer to promote the metallization of microencapsulated paraffins. Zimmerer et al. [165] developed an eco-friendly technique for creating phase change material encapsulated in a thin (50 nm) nickel coating using electroless metallization on PDA-modified PCMs, which can be used for heat conversion, storage, and protecting temperature-sensitive components (Fig. 4.4d). In this work, an ultra-thin and smooth PDA layer was added, which was suitable for the subsequent metallization. The coated microcapsules demonstrated a phase change enthalpy of around 170 J/g, making them suitable for thermal applications. The nickel coating also improved the thermal diffusivity of the microcapsule powders and allowed rapid microwave heating, demonstrating their potential for controlled heating uses.

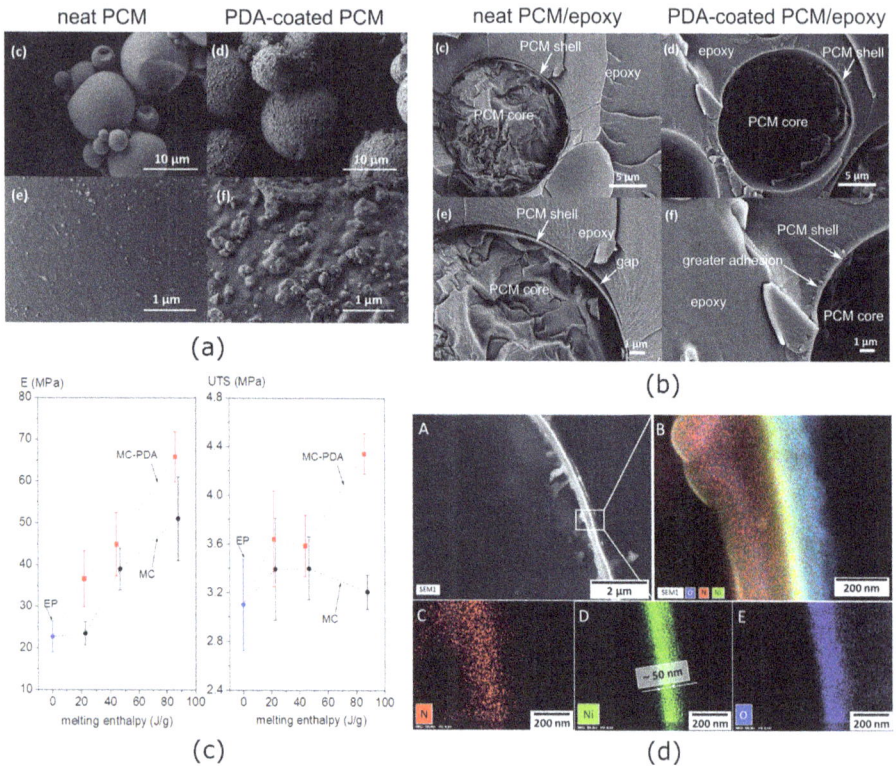

Fig. 4.4: (a) Neat and PDA-coated PCM microcapsules (duplicated from [163]). (b) Neat and PDA-coated PCM microcapsules embedded in an epoxy matrix. The greater adhesion due to PDA coating is evidenced (adapted with permission from [163]). (c) Elastic modulus (E) and ultimate tensile strength (UTS) of flexible epoxy/PCM composites with neat (MC) or PDA modified (MC-PDA) microcapsules (duplicated from [164]). (d) SEM images of the MC shell coated with PDA and Ni, with the corresponding colored elemental maps of N, Ni, and O. The elements, highlighted in different colors, help visualize especially the thickness of the Ni coating (duplicated from [165]).

4.3.2 PCMs in structural polymer composites

Excluding the construction materials field, there is a surprisingly limited number of publications about materials designed to combine structural and TES functions. One of the first reported examples is that presented by Wirtz et al. [166], who developed a sandwich structure made of a graphitic foam impregnated with paraffin (= 56 °C) as the core and carbon/epoxy laminates as the skin, intended for the thermal management of electronic devices. The graphitic foam had the twofold function of immobilizing the molten paraffin and enhancing the thermal conductivity, while the carbon/epoxy skins increased the flexural stiffness. The higher the foam porosity, the higher the paraffin amount that could be accommodated into the foam pores, and the lower its mechanical properties. Therefore, the foam porosity was the main parameter dominating the balance between the core's mechanical properties and the overall TES performance, which acted competitively also in this system, even though the carbon-fiber-based skins were the main enhancers of the mechanical performance of the whole sandwich structure. The literature reports several other studies on carbon foams as shape-stabilizers for organic PCMs, but they are generally employed only to enhance the thermal conductivity, while little attention is given to the mechanical performance [167–169].

More recently, a research group at RMIT University (Melbourne, Australia) developed a woven glass fiber/epoxy laminate containing paraffin microcapsules and investigated the impact of the PCM on the mechanical and thermal properties of the host laminate [170–172]. An increase in the PCM weight fraction determined a decrease in the tensile, flexural, and compressive properties, due to the intervention of additional damaging mechanisms such as delamination, matrix cracking, and fiber/ matrix debonding, and the authors indicated the weak interfacial adhesion between the capsules and the matrix as one of the causes for the mechanical performance decrease. On the other hand, the mild processing conditions allowed for retaining most of the initial PCM enthalpy, but thermogravimetric analysis (TGA) showed that a fraction of the capsules was damaged during processing. Nevertheless, these results were promising for the development of such structural TES polymer-matrix composites, which exhibited a phase change enthalpy of up to 40 J/g.

The following subsections report some other examples and case studies of thermoplastic and thermosetting polymer composites containing different types of phase change materials. The selected case studies aim to highlight key aspects of the production process and the final thermal and mechanical performance of the produced composites. More specifically, it will be evidenced how the production process varies depending on the selected matrix, how the process parameters affect the effective PCM concentration in the final composite, and how the type and amount of PCM alter the microstructure and the thermomechanical and heat storage performance of the host composite.

4.3.2.1 Case study 1: impact of microencapsulated paraffin on the thermomechanical properties of a unidirectional carbon fiber/epoxy composite

This case study [173] explores the production of structural carbon fiber/epoxy composites with thermal management capabilities through the incorporation of microencapsulated paraffin. While other research has investigated various PCM-enhanced polymer composites, the unique aspect of this case study lies in its focus on unidirectional carbon/epoxy laminates, which allows for a clearer understanding of how PCM addition affects fiber-related and matrix-related properties separately. The composites were produced via a mild production process, which preserved the integrity of the PCM microcapsules while ensuring uniform distribution throughout the composite. Commercial melamine-formaldehyde encapsulated paraffin ($T_m = 43$ °C) with microcapsule diameters ranging from 15 to 30 μm was employed. The production method involved first dispersing different concentrations of PCM microcapsules (20, 30, and 40 wt% of the total weight of the matrix) into an epoxy matrix through manual stirring, followed by a conventional wet lay-up technique using unidirectional carbon fiber fabric (150 g/m^2). Two types of laminates were produced: 8-ply laminates for general characterization and 16-ply laminates specifically designed for interlaminar testing, with the latter incorporating a PET film in the mid-plane to generate a pre-crack. The manufacturing process concluded with room temperature curing for 24 h followed by post-curing at 100 °C for 10 h under vacuum bag conditions. This method proved effective in producing composite laminates with thicknesses ranging from approximately 1.3 mm for the neat samples to 2.4 mm for the highest PCM content in the 8-ply configuration, and from 3.0 mm to 5.6 mm for the 16-ply variants.

The characterization of these composites encompassed microstructural, thermal, dynamic mechanical, and mechanical characterization. Microstructural analysis through optical microscopy revealed that, in the reference epoxy-carbon fiber laminate, fibers exhibited uniform distribution throughout the thickness, while the PCM-containing laminates showed preferential accumulation of microcapsules in the interlaminar regions rather than within individual fiber tows (Fig. 4.5), attributable to the size differential between the microcapsules (20 μm average diameter) and carbon fibers (7 μm average diameter). The thickening of the interlaminar region caused the fiber volume fraction to drop. In fact, thermogravimetric analysis enabled the determination of the actual fiber content, revealing a systematic decrease with increasing PCM concentration. The fiber weight fraction dropped from 71.5% in the reference laminate to 41.8% in the highest PCM content variant, indicating that higher PCM loadings hindered matrix flow during processing.

On the other hand, the PCMs positively contributed to the thermal management capability of the prepared composites. DSC measurements demonstrated that all composites maintained their intended thermal storage functionality. The PCM phase transitions occurred between 40–60 °C for melting and 40–15 °C for crystallization, with the glass transition temperature of the epoxy matrix slightly increasing with PCM content. The energy storage capacity scaled proportionally with PCM content, reaching maximum values of 48.7 J/g for composites containing 40 wt% microcapsules in the initial mixture. The TES effect was captured on a larger scale via thermal imaging tests, which demonstrated the practical thermal management capabilities of the composites. For instance, while the reference laminate reached 60 °C in 2.3 min during heating, the composite with 40 wt% PCM required 8.2 min, illustrating effective thermal buffering. Similar behavior was observed during cooling, with the time to reach 30 °C extending from 3.4 to 15.8 min between the reference and highest PCM content samples (Fig. 4.6).

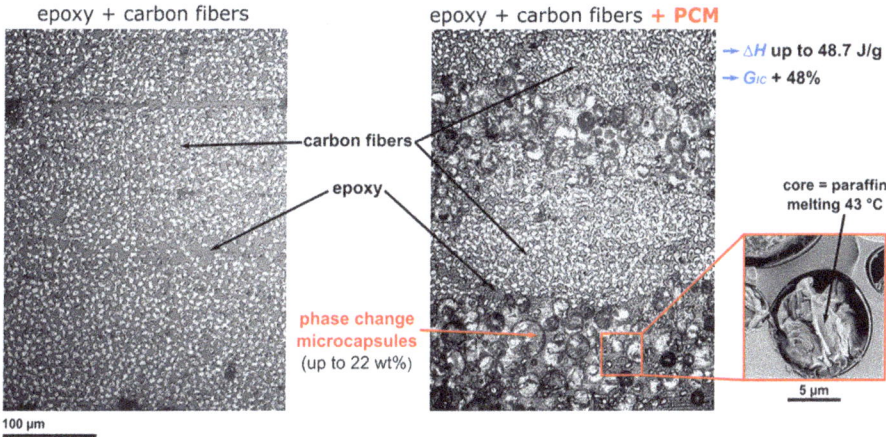

Fig. 4.5: Optical microscope micrographs of a neat carbon fiber/epoxy laminate (left) and a PCM-added carbon fiber/epoxy laminate with evidence of PCM concentration in the interlaminar region (duplicated with permission from [173]).

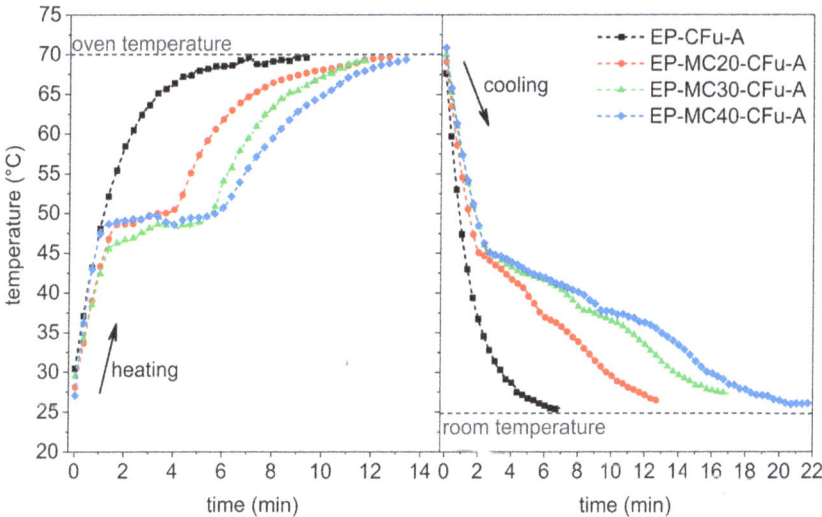

Fig. 4.6: Surface temperature of the laminates without (EP-CFu-A) and with increasing PCM concentrations as a function of the testing time during heating and cooling, as measured with a thermal imaging camera (duplicated with permission from [173]).

Interestingly, these composites were also subjected to dynamic-mechanical thermal analysis (DMTA), which revealed significant insights into their viscoelastic behavior across temperature ranges encompassing both the PCM phase transition and the matrix glass transition. The storage modulus (E') exhibited a distinctive two-step decrease: the first corresponding to PCM melting (5–45 °C) and the second to the epoxy matrix glass transition (around 100 °C). The magnitude of the modulus drops at PCM melting showed a strong linear correlation with PCM content, with R^2 values of 0.998. Cyclic DMTA testing demonstrated good recovery of mechanical properties through melting-crystallization cycles. After thermal cycling between –40 °C and 60 °C, the storage modulus recovered to approximately 95% of its initial value for PCM-containing samples, compared to 99.6% for the reference laminate. This behavior indicated good structural stability during thermal cycling, though with a slight hysteresis effect evidenced by crystallization peaks occurring at lower temperatures than melting peaks. Finally, multi-frequency DMTA testing (0.3–30 Hz) provided insights into the temperature-frequency dependence of the viscoelastic response. The activation energy for the glass transition, calculated using an Arrhenius approach, increased with PCM content from 365 kJ/mol in the reference laminate to 419–434 kJ/mol in PCM-modified samples, suggesting that PCM addition might restrict polymer chain mobility at the glass transition. An innovative aspect of the presented DMTA characterization was the attempt to analyze the frequency dependence of the PCM melting transition. While more scattered than glass transition data, the analysis yielded activation energies for melting ranging from 615 kJ/mol to 955 kJ/mol for increasing PCM contents, though with larger statistical uncertainty ($R^2 \approx 0.96$–0.97). Hence, the DMTA analysis proved particularly valuable for understanding the composite behavior across operating temperatures, demonstrating that while PCM incorporation introduces additional thermal transitions, the materials maintain consistent mechanical performance through thermal cycling. This characterization approach established new methodologies for evaluating phase-change phenomena through dynamic mechanical testing.

The incorporation of microencapsulated paraffin significantly influenced the mechanical properties of the carbon fiber/epoxy laminates across multiple testing modes. The tensile response in the fiber direction (longitudinal) showed a systematic decrease in modulus with increasing PCM content. As demonstrated by theoretical modeling using mixture rules, this reduction in stiffness was predominantly due to the decreased fiber content rather than any inherent weakening [174] of the matrix by the PCM inclusions. The flexural behavior revealed distinctive failure mechanisms between the reference and PCM-modified composites. While the unmodified laminates exhibited catastrophic failure originating from the tensile region, PCM-containing samples showed progressive failure with characteristic load drops and plateaus, indicating enhanced energy dissipation during damage progression. The flexural modulus decreased with PCM content, following the trend observed in tensile properties. A notable transition in failure mode was observed, shifting from fiber-dominated failure in neat laminates to interlaminar-dominated failure in PCM-modified composites. The flexural test was also performed at 65 °C, i.e., above the PCM melting point, at which the flexural modulus remained relatively stable for low PCM contents, though moderate decreases emerged at higher concentrations. Instead, strength retention varied with PCM content, maintaining 74% of room temperature values at 20 wt% PCM but declining to 59% as PCM content increased to 40 wt%. Similarly, strain-to-failure values showed temperature sensitivity, retaining between 70% and 79% of their room temperature values across the range of PCM concentrations.

A very interesting test was the evaluation of the mode I interlaminar fracture toughness through the double cantilever beam test. This property was maximum at moderate PCM concentrations, with an initial PCM addition of 20 wt% increasing the critical strain energy release rate for both initiation (G_{II}) and propagation (G_{IC}). The initiation toughness improved markedly from 0.15 kJ/m^2 to 0.34 kJ/m^2 with a 20 wt% PCM addition. However, higher PCM loadings led to declining toughness values, with G_{Ic} decreasing to 0.30 kJ/m^2 at 40 wt% PCM. This transition in the toughening mechanism was evidenced by the loss of fiber bridging at higher PCM contents, indicating a fundamental change in crack propagation behavior.

Hence, the integration of microencapsulated paraffin into structural carbon fiber composites demonstrates promising multifunctional performance, particularly at moderate PCM contents. The achieved thermal storage capacity of 48.7 J/g, combined with maintained mechanical integrity, suggests viable applications in temperature-sensitive structural components. However, the decreased fiber volume fraction with PCM addition remains a key challenge affecting mechanical properties. Future development should address three critical aspects: processing optimization to maintain higher fiber volume fractions, enhancement of PCM-matrix interfacial properties, and a deeper understanding of micromechanical behavior. Processing strategies should focus on managing matrix viscosity and developing improved cure cycles. Interfacial engineering through surface modification of microcapsules and investigation of specific coupling agents could enhance mechanical properties, particularly in the interlaminar regions where PCMs tend to accumulate.

4.3.2.2 Case study 2: polyamide 12/discontinuous carbon fiber thermoplastic composites containing a microencapsulated phase change material

This case study [175] involves the production of thermoplastic composites containing discontinuous carbon fibers and microencapsulated phase change materials. Polyamide 12 (PA12) was selected as the matrix material due to its favorable processability, mechanical properties, and compatibility with the microencapsulated paraffin wax PCM. The composites were fabricated via melt compounding and hot plate pressing. The PA12 matrix was blended with varying loadings of the PCM microcapsules (20–60 wt%) and two types of carbon fibers – chopped (CFL, length 6 mm) and milled (CFS, length 100 μm) – at 10–30 wt% loading. The melt compounding step was critical to ensure homogeneous dispersion of the PCM capsules and carbon fibers within the PA12 matrix. Care was taken during this high-shear process to minimize damage to the PCM microcapsules, which could reduce their thermal energy storage capacity. The compounded materials were then compression molded into square sheets for subsequent characterization.

SEM micrographs (Fig. 4.7a) revealed a homogeneous dispersion of both the long (CFL) and short (CFS) carbon fibers within the PA12 matrix, along with good fiber-matrix adhesion, particularly for the longer CFL fibers. The PCM microcapsules were also well distributed, though some debonding and capsule breakage were observed, likely due to the high shear stresses during melt compounding. DSC showed that the melting/crystallization enthalpy values increase with the PCM weight fraction up to 60 J/g. Although the phase change temperatures of the paraffin were unaffected by the composite composition, the measured latent heat of fusion was lower than expected, especially at higher PCM loadings and for the CFL-containing composites. However, the relative phase change enthalpy, calculated as a percentage of the expected value based on the PCM content, was 41–94% of the expected value and decreased with an increase in the fiber content. This reduction in relative enthalpy was more pronounced for the composites containing longer CFL fibers compared to those with the shorter CFS fibers, suggesting more severe damage to the PCM microcapsules and paraffin leakage during the high-shear melt compounding process. This was confirmed by dynamic rheological testing, which showed that the addition of fillers, particularly the longer CFL fibers, significantly increased the melt viscosity, potentially contributing to more severe damage to the PCM microcapsules.

Tensile testing (Fig. 4.7b) indicated that the incorporation of PCM microcapsules decreased the mechanical properties, such as elastic modulus, tensile strength, and ductility, while the addition of carbon fibers improved not only the stiffness and strength, but also the thermal stability, diffusivity, and conductivity of the composites, which is beneficial for enhancing the thermal energy storage and release performance. More specifically, the elastic modulus of the neat PA12 matrix was 1.20 ± 0.07 GPa, and this value decreased to 0.59 ± 0.07 GPa for the composite containing the highest PCM content of 60 wt%. The inclusion of carbon fibers helped mitigate this reduction, with the composite containing 20 wt% CFL exhibiting an elastic modulus of 1.81 GPa. However, the strain at break was significantly lowered by the addition of both the PCM microcapsules and the carbon fibers.

The employed processing parameters were milder than those utilized to produce continuous glass fiber laminates having the same PA12 matrix and the same PCM [176], for which even greater PCM degradation and leakage were experienced. However, to further optimize the performance of these multifunctional composites, future work should focus on further minimizing damage to the PCM microcapsules during processing, as well as exploring alternative fiber types or surface treatments to improve the fiber-matrix adhesion and reduce fiber length decrease and investigating the use of other matrix materials or hybrid filler systems to achieve an optimal balance of thermal and mechanical properties.

(a)

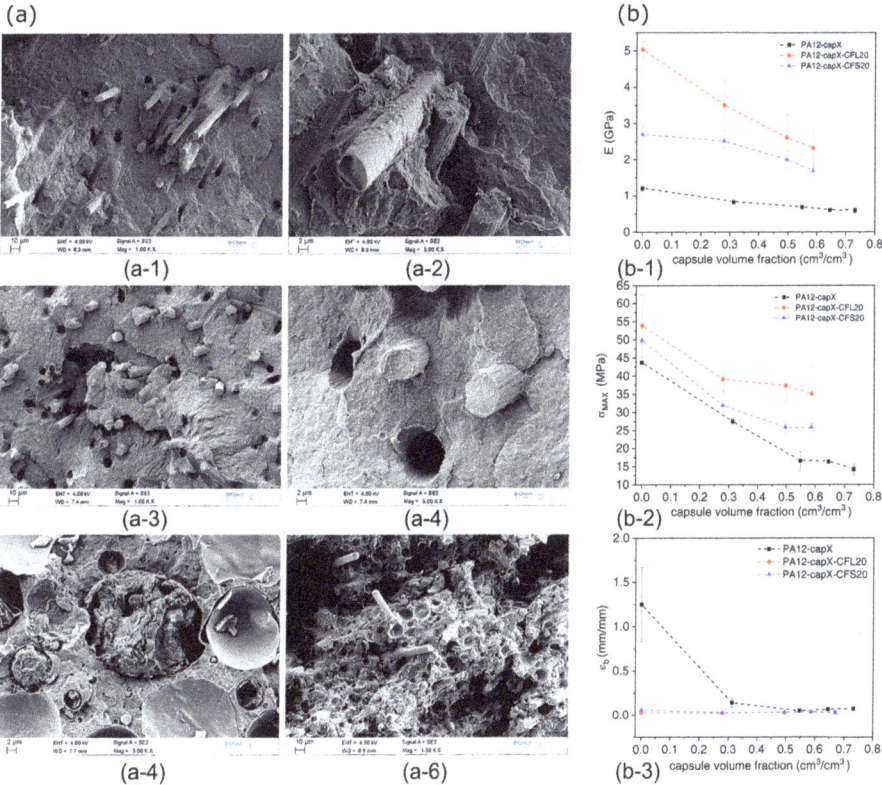

(b)

(a-1)

(a-2)

(b-1)

(a-3)

(a-4)

(b-2)

(a-4)

(a-6)

(b-3)

Fig. 4.7: (a) SEM micrographs of the cryofracture surface of the PA12/CF/PCM composites containing (a-1/2) 20 wt% CFL; (a-3/4) 20 wt% CFS; (a-5) 50 wt% PCM; (a-6) 50 wt% PCM compared to the total matrix / PA12 + PCM concentration) and 20 wt% CFL. (b) Results of the mechanical characterization of the prepared composites (b-1) elastic modulus, (b-2) tensile strength, and (b-3) strain at break as a function of the PCM ("cap") volume fraction. Adapted with permission from [175].

4.3.2.3 Case study 3: ultrathin wood laminae/thermoplastic starch biodegradable composites containing polyethylene glycol as the PCM

This case study [177] presents an innovative approach to creating fully biodegradable multifunctional structural TES composites. The work addresses growing environmental concerns about traditional polymer matrix composites by utilizing entirely biodegradable constituents: beech wood laminae as the reinforcement, biobased thermoplastic starch (TPS) as the matrix, and PEG (MW 2000 g/mol, T_m = 51–55 °C) as the PCM. The particular interest in this system stems from the potential of wood's rough and porous microstructure to act as a natural container for PCM stabilization, while simultaneously serving as a structural reinforcement. This approach offers a sustainable alternative to conventional composites while adding thermal management functionality, making it particularly attractive for applications in green building materials, sustainable packaging with temperature control, and eco-friendly thermal insulation systems.

The composites were manufactured through a two-step process. First, ultrathin beech laminae (320 μm thickness) were impregnated with molten PEG at 70 °C for 5 min, followed by gentle blotting to remove excess PCM from the surface. The impregnated laminae were then alternated with thermoplastic starch sheets (100 μm thickness) and hot-pressed at 170 °C for 7 min under a pressure of 1 MPa. Each laminate consisted of 5 wood/PEG laminae alternated with 6 starch sheets. Two different stacking sequences were investigated: unidirectional [0_5] (tested both along the fiber direction (L) and in the transversal direction (T)) and angle-ply [+45/-45/+45/-45/+45] (Fig. 4.8a). This processing approach notably avoided the need for a melt compounding step that could potentially degrade the PCM, thereby preserving its thermal energy storage properties.

The characterization revealed several interesting findings. While the initial PEG absorption by wood laminae was substantial (86 wt% of the wood's initial mass), about 25% of the total mass was lost during hot-pressing, resulting in an effective PCM content of approximately 11 wt% in the final composite. SEM analysis showed partial filling of wood porosity by PEG, though the PCM presence on laminae surfaces somewhat compromised matrix-reinforcement adhesion (Fig. 4.8b).

DSC measurements confirmed successful PCM incorporation, with the composites exhibiting a melting enthalpy of 27.4 J/g at 55 °C. Notably, contrary to typical PCM-containing composites where PCM addition often deteriorates mechanical properties, these composites showed enhanced performance. Both tensile and impact properties were improved compared to composites not containing the PCM, which was attributed to the partial filling of wood porosity by PEG. For example, in the L direction, the tensile modulus increased from approximately 1 GPa to approximately 7 GPa, the tensile stress at break increased from approximately 20 MPa to approximately 45 MPa, and notably the impact energy increased from 1.2 J/m^2 to 23.7 J/m^2.

This case study demonstrates the feasibility of creating fully biodegradable multifunctional composites with synergistic rather than parasitic effects between thermal and mechanical properties. However, several aspects could be improved in future developments. The significant PCM leakage during processing suggests the need for better PCM retention strategies, perhaps through chemical modification of the wood surface or optimization of the impregnation process. The matrix-reinforcement interface could be enhanced through surface treatments or coupling agents compatible with the system's biodegradability requirements. Additionally, future work could explore the incorporation of thermally conductive natural fillers to enhance heat transfer within the composite, as well as investigate the long-term stability of the PCM within the wood structure under thermal cycling. The scalability of the manufacturing process and its adaptation to industrial production methods also warrant further investigation [177].

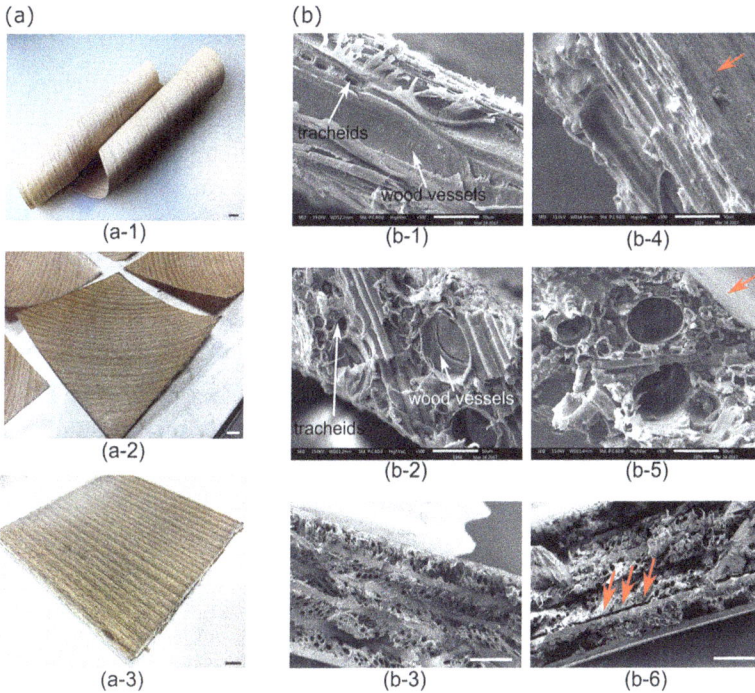

Fig. 4.8: (a) Representative images of (a-1) neat wood lamina; (a-2) wood laminae after PEG impregnation; (a-3) wood/PEG/starch laminate. (b) SEM micrographs of the cryofracture surface (cross-section) of (b-1) neat wood lamina, longitudinal direction; (b-2) neat wood lamina, transversal direction; direction; (b-2) neat wood/starch laminate, transversal direction; (b-4) wood/PEG lamina, longitudinal direction (arrows indicate PEG on the lamina surface) (b-5) wood/PEG lamina, transversal direction (arrows indicate PEG on the lamina surface); (b-6) wood/PEG /starch laminate, transversal direction (arrows indicate delamination). Scale bars: (a) 1 cm; (b-1/2/4/5) 50 μm; (b-3/6) 500 μm. Adapted from [177].

4.3.2.4 Case study 4: multifunctional sandwich composites with a PCM-enriched polyurethane foam core and epoxy/carbon fiber skins

The last of the presented case studies deals with the production and characterization of sandwich composites made of polyurethane/PCM foam cores and high-performance epoxy/carbon fiber structural skins [178]. These composites are designed for use in cold chain logistics applications, where the integration of PCM-enhanced cores into sandwich structures enables simultaneous thermal regulation and load-bearing performance. By optimizing the PCM content within the PU foam core, this composite aims to strike an optimal balance between mechanical properties, crucial for structural integrity during shipping and handling, and thermal energy storage capabilities to maintain product temperature during transport and storage. The strategic use of PCM-enhanced sandwich composites in refrigerated transportation and storage can facilitate significant mass and volume savings compared to separate structural and thermal management systems, making them an attractive solution for enhancing the energy efficiency and sustainability of cold chain logistics.

The fabrication of the multifunctional sandwich composites involved a two-step process. First, PU foam cores containing varying amounts of microencapsulated PCM ($T_m = 32$ °C) were produced and characterized to identify the optimal PCM content. The PU foam cores were prepared by mixing polyol and isocyanate precursors, with the PCM incorporated at 10, 20, or 30 wt% of the total foam composition. The foam samples were cast into molds and cured at 40 °C.

The characterization of the PCM-enriched foam cores provided important insights into the impact of PCM incorporation on the microstructure, thermal properties, and mechanical performance. The microstructural analysis revealed that as the PCM content increased, the cell structure transitioned from a typical closed-cell polyhedral configuration to a more open-cell morphology with larger, spherical cells. This cell opening effect was attributed to the PCM microcapsules preferentially located at the cell strut interfaces and promoting their rupture. The increased open porosity and reduced closed porosity with higher PCM loadings led to a rise in both the apparent and bulk densities of the foams (Fig. 4.9). Thermal analysis by DSC confirmed the ability of the PCM to provide latent heat storage, with the stored energy increasing linearly with PCM content up to 42 J/g for the 30 wt% PCM-containing foam. However, the relative enthalpy values indicated that only 80% of the theoretical PCM was effectively participating in the phase change. Thermal conductivity testing showed that the increase in PCM content from 0 to 30 wt% raised the thermal conductivity from 0.024 to 0.037 W/(m · K), though the foams remained within the typical insulation range. Mechanical characterization demonstrated that the flexural properties were less impacted by PCM addition compared to the compressive properties, with the 20 wt% PCM composition identified as the optimal balance between thermal energy storage, thermal insulation, and mechanical performance for use in the sandwich composite panels.

Hence, in the second step, sandwich panels were fabricated by using either the neat PU (S-PU) or the PU + 20%PCM foam as the core (S-PU-PCM20), with high-performance epoxy/carbon fiber laminates as the skins. The sandwich panels were produced via hand layup and vacuum bagging, ensuring good interfacial adhesion between the core and skins. The fabricated sandwich panels underwent comprehensive microstructural, thermal, and mechanical characterization to evaluate the effects of PCM incorporation on the overall multifunctional performance.

Microstructural analysis using light microscopy (Fig. 4.10) revealed a strong and defect-free interface between the foam cores and the epoxy/carbon fiber laminate skins, even in the PCM-containing samples. This indicated that the addition of PCM microcapsules did not impair the interfacial adhesion, likely due to the effectiveness of the epoxy resin infiltration during the hand layup process.

Thermal conductivity measurements showed that the sandwich panels preserved the thermal insulation properties of the foam cores, with only a modest increase in

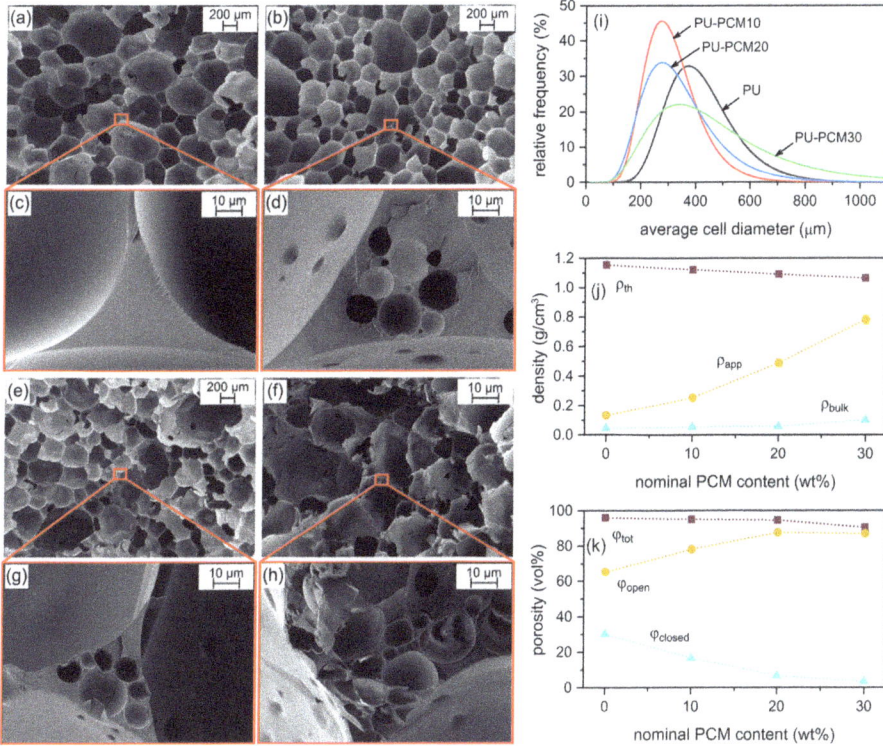

Fig. 4.9: Microstructural and physical characterization of the prepared foams. (a–h) SEM micrographs of the cryofracture surface of the samples (a,c) PU; (b,d) PU + 10 wt% PCM (PU-PCM10); (e,g) PU + 20 wt% PCM (PU-PCM20); (f,h) PU + 30 wt% PCM (PU-PCM30); (i) Log-normal fit of the average foam cell size distribution obtained from the analysis of the SEM micrographs of foam samples; (j) theoretical, apparent, and bulk density of the prepared foams as a function of the PCM concentration; (k) total, open, and closed porosity of the prepared foams as a function of the PCM concentration (duplicated from [178]).

thermal conductivity observed for the PCM-containing samples compared to the neat PU cores (34 vs. 36 mW/(m · K) at 30 °C). Mechanical testing revealed some differences in the failure modes and strengths between the neat PU and PCM-containing sandwich panels (Fig. 4.11). While the flexural properties were relatively unaffected by PCM addition, the edgewise compressive strength and flatwise tensile strength were moderately reduced in the PCM-containing samples. This was attributed to the lower compressive and tensile properties of the PCM-enhanced foam cores compared to the neat PU foam. Nevertheless, the sandwich panels maintained adequate mechanical performance for the targeted cold chain logistics applications.

In conclusion, this case study demonstrated the promising potential of multifunctional sandwich composites that integrate phase change materials within their foam cores to simultaneously provide structural integrity and thermal regulation capabilities. The systematic investigation of the PU/PCM foam cores revealed an optimal bal-

Fig. 4.10: Light microscope images of the polished cross-section of the sandwich samples, highlighting the interfacial adhesion. Low (a) and high (c) magnification micrograph of S-PU sample. Red arrows indicate porosity in the epoxy/CF laminate; low (b) and high (d) magnification micrograph of the S-PU-PCM20 sample (duplicated from [178]).

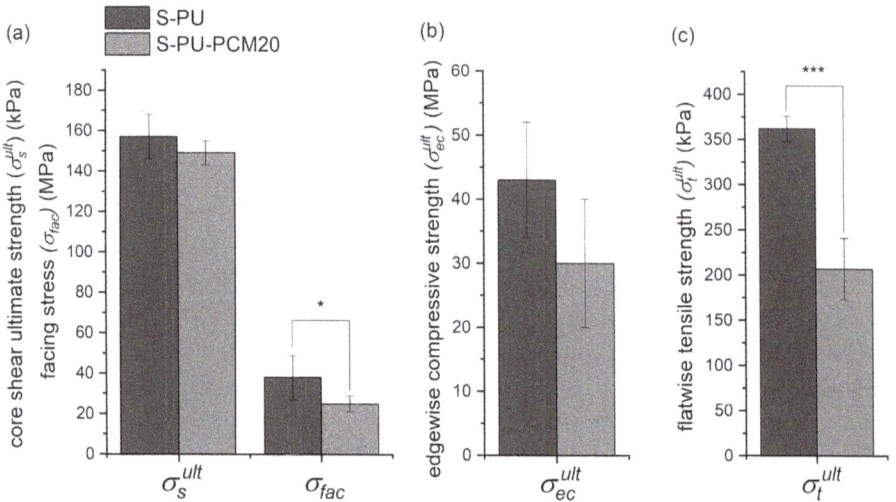

Fig. 4.11: Summary of the results of the mechanical properties of the prepared sandwich panels. (a) results of the three-point bending tests (σ_s^{ult} = core shear ultimate strength; σ_{fac} = facing stress); (b) results of the edgewise compression tests (σ_{ec}^{ult} = ultimate edgewise compressive strength); (c) results of the flatwise tensile tests (σ_t^{ult} = ultimate flatwise tensile strength). Asterisks indicate statistically significant differences according to the one-way ANOVA test (p-value = 0 "****" 0.001 "***" 0.01 "**" 0.05). Duplicated from [178].

ance between thermal energy storage and mechanical performance at 20 wt% PCM content, a formulation that was then successfully incorporated into the final sandwich panel designs. While the incorporation of PCM did lead to a moderate reduction in certain mechanical properties, such as edgewise compressive and flatwise tensile strengths, the overall structural performance remained adequate for the target applications in refrigerated transportation and storage. Moving forward, further optimization of the foam core composition and microstructure, potentially through the use of additives or alternative PCM types, presents opportunities to enhance the shear resistance and mitigate the trade-offs in mechanical performance. Additionally, scale-up testing under real-world conditions will be crucial to validate the thermal regulation capabilities of these multifunctional composites and pave the way for their widespread adoption in energy-efficient and sustainable cold chain logistics applications.

5 Evaluating the multifunctional performance of structural TES composites

From the case studies presented in Section 4.3.2, and also from the rest of the scientific literature on structural TES composites, one can conclude that the introduction of an organic PCM in a fiber-reinforced composite does increase its TES capability and thermal management properties, but it often impairs its mechanical performance, especially at high PCM weight fractions. In other words, despite the potential of such materials, the structural and TES properties are hardly ever synergistic. This occurs for three main reasons. The first is that the addition of a third phase unavoidably decreases the maximum fiber volume fraction, given that the matrix volume fraction cannot decrease under a certain threshold. The second is that commercial PCMs are generally not intended as fillers for polymer matrices, and surely they are never aimed at improving the mechanical properties of the host matrices (i.e., at being used also as a reinforcing phase). Therefore, the thermal, mechanical, and surface properties of such PCMs are not optimized for this purpose. The third reason is that, on the other hand, the reinforcement hardly ever acts as the shape-stabilizing phase for the PCM. If it did, the introduction of the PCM in the composite would be "volumeless," thus leaving the mechanical properties unaltered. This is more or less what has been observed in Case Study 3, where the ultrathin wood laminae contributed to the shape-stabilization of PEG, and PEG contributed to the mechanical properties of the laminate (yet only below its melting point). However, this case is an exception, while the situation described in the other reported examples is much less enthusiastic.

Nevertheless, combining structural and TES functions in a composite material may still be advantageous in terms of mass and volume savings, compared to two monofunctional units performing the structural and TES functions individually [166]. To quantify this advantage, it is important to develop objective selection and design criteria that consider all the multiple and sometimes competing design requirements of such structural TES composites.

Hence, this chapter aims to describe a parameter to quantitatively evaluate the multifunctionality of composites that perform both the structural and TES functions. Specifically, by following the approach first described in [179], a criterion is discussed to minimize the mass of a component with both thermal energy storage and structural functions, and then a multifunctionality parameter is developed that quantifies the mass saving at the system level. This parameter is then applied to evaluate and rank some of the examples of structural TES composites found in the literature. This approach allows the development of objective design principles and material selection guidelines, fundamental to maximizing the advantages of using a multifunctional material.

https://doi.org/10.1515/9783111111865-005

5.1 Relationship between the structural and TES performance: synergistic or parasitic?

The scientific literature contains several examples of structural TES composites with a wide range of polymer matrices, reinforcements, and PCMs. The works encompassed (i) thermoplastic, thermosetting, and reactive thermoplastic matrices, (ii) traditional fibrous reinforcements constituted by continuous or discontinuous glass and carbon fibers and less traditional reinforcements such as thin beechwood laminae, and (iii) microencapsulated and shape-stabilized PCMs, added in variable weight fractions (up to ~ 30 wt%). All the considered systems are listed in Tab. 5.1.

Tab. 5.1: Structural TES polymer composites under investigation in this chapter with specifications of matrices, reinforcing agents, and PCMs. Labels help interpret Figures 5.1 and 5.3 ("x" indicates the weight fraction of PCM in the matrix or, just for the systems EP-GFx, the approximate volume fraction of PCM in the composite; "y" indicates the weight fraction of the fibers in the whole composite).

Matrix	Reinforcement	PCM	Label	Ref.
Polyamide 12	Bidirectional glass fibers	CNT-stabilized paraffin	PA-ParCNTx-GF	[156, 176]
Polyamide 12	Bidirectional glass fibers	Paraffin microcapsules	PA-MCx-GF	[176]
Reactive acrylic thermoplastic (Elium)	Bidirectional carbon fibers	Paraffin microcapsules	EL-MCx-CF	[180, 181]
Thermoplastic starch	Thin beechwood laminae	PEG 600	wood/starch/PEG	[177]
Epoxy	Bidirectional carbon fibers	CNT-stabilized paraffin	EP-ParCNTx-CF	[156, 182]
Epoxy	Unidirectional carbon fibers	Paraffin microcapsules	EP-MCx-CFu	[173]
Epoxy	Glass fibers, cross ply	Paraffin microcapsules	EP-GFx	[170, 171]
Epoxy	Milled carbon fibers	Paraffin microcapsules	EP-MCx-CFSy	[174, 183]
Polyamide 12	Chopped (CFL) or milled (CFS) carbon fibers	Paraffin microcapsules	PA-MC-CFSy; PA-MCx-CFLy	[175]

To investigate the relationship between structural and TES performance and compare the properties of the different composites combining structural and TES functions found in the literature, their elastic modulus and mechanical strength (structural parameters) can be plotted as a function of their melting enthalpy (TES parameter). The results can be observed in Fig. 5.1.

The elastic modulus (Fig. 5.1a) generally decreases with increasing melting enthalpy. For continuous-fiber composites, the decrease is mainly due to a drop in the reinforcement volume fraction, as is also clear considering that the decrease is almost negligible when the fiber volume fraction has been forced constant by adjusting the processing parameters, such as in the laminates epoxy/CF/paraffin + CNTs (0–90) [182]. Similar conclusions can be drawn from the results of discontinuous fiber composites, for which the fiber volume fraction is generally lower and easy to control and the modulus is significantly dependent also on the matrix stiffness. In this case, when the PCM fraction is constant, the melting enthalpy is constant and the modulus increases with the fiber content; on the other hand, when the fiber fraction is constant, the melting enthalpy increases and the elastic modulus slightly decreases with the PCM content. Similar considerations can be made for the mechanical strength (Fig. 5.1b), which generally decreases with increasing phase change enthalpy.

In most of the investigated applications, the ideal case would be a combination of high stiffness (and strength) and high melting enthalpy, which would be represented in the top right corner of Fig. 5.1a–b. However, it is very challenging to maximize both properties simultaneously, because the mechanical properties generally increase with the fiber volume fraction and the TES properties with the PCM weight fraction, while the experimental results show that fiber and PCM fractions generally follow opposite trends. The best materials for such composites would be a reinforcement with high stiffness-to-density and stress-to-density ratios and a PCM with a large phase change enthalpy-to-specific volume ratio, so that the product between fiber volume fraction and PCM mass fraction would be maximized. In any case, the property to be maximized depends on the specific application, as well as the combination of properties to be considered optimal.

The trends depicted in Fig. 5.1 occur because, for most of the studied systems, the reinforcement does not contribute to store heat and the PCM does not raise the stiffness and strength. To make structural and TES properties truly synergistic and not parasitic, the multifunctionality should be shifted from the level of the composite to the level of the single phase. This was achieved in one of the investigated systems, i.e., the wood/TPS/PEG laminate. In this composite, thin beechwood laminae were impregnated with PEG, interleaved with thermoplastic starch sheets, and compacted by hot pressing. Here, the beechwood laminae acted both as reinforcement and shape-stabilizing agent for PEG, and PEG not only played the role of the PCM, but also significantly increased the tensile, impact, and dynamic-mechanical properties of the composite. In fact, the wood/TPS/PEG laminate is the only example in which the mechanical properties do not inversely correlate with the melting enthalpy.

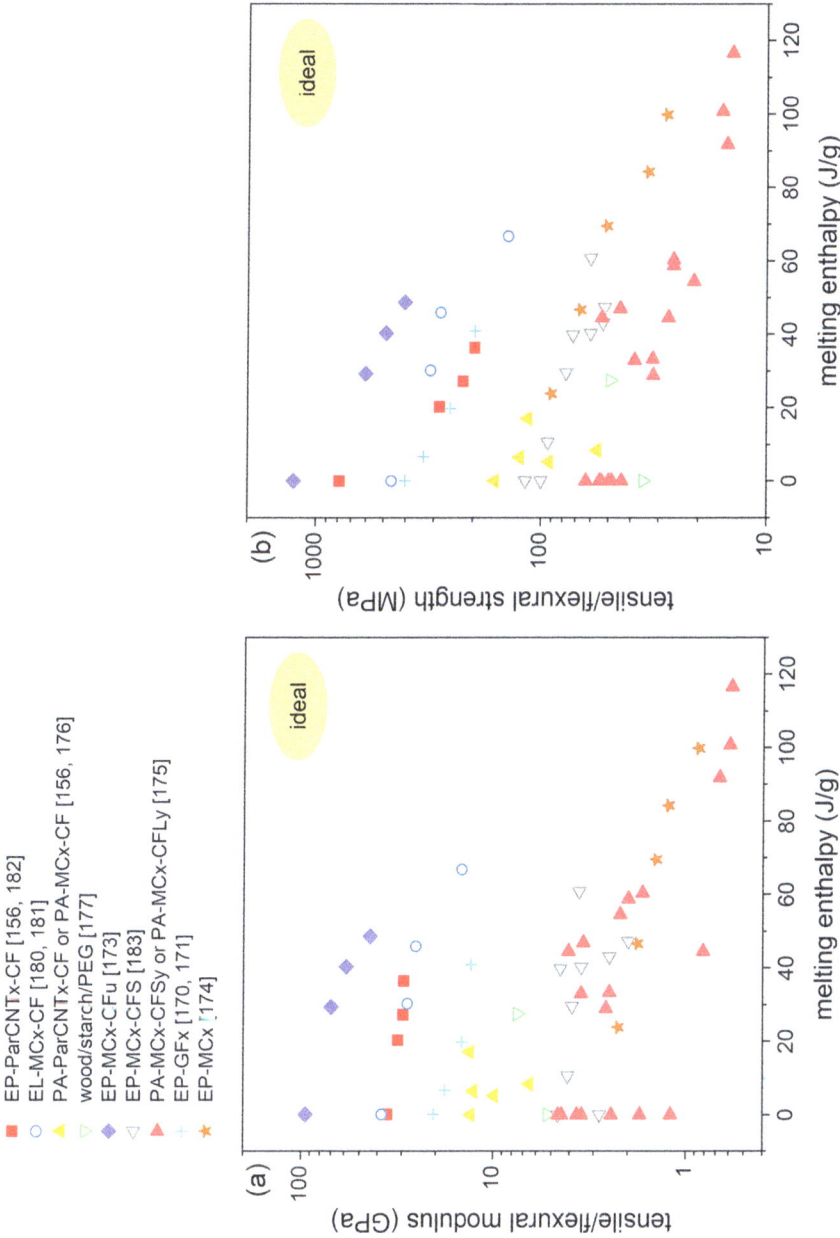

Fig. 5.1: Mechanical properties as a function of the melting enthalpy for examples of structural TES composites found in the literature. (a) Elastic modulus; (b) Mechanical strength. A general decrease in the mechanical properties is evidenced by an increase in the TES capabilities. However, the ideal condition is the one with high mechanical properties and high TES capabilities (top right corner of the graphs). Values of properties along the fiber axis for unidirectional (UD) composites and in the 0–90 direction for bidirectional (plain weave, 0–90 or cross-ply, CP) composites are reported.

5.2 Determination of the multifunctional efficiency

To reconcile the competing design requirements of the presented multifunctional structural TES composites, it is important to define objective parameters that encompass all functions of the investigated material. The approach discussed here, presented for the first time in [179], is similar to that developed by O'Brien et al. [184] for multifunctional structural composite capacitors. The objective of this approach is to minimize the system mass for a unit presenting both structural and TES constraints.

The approach involves the comparison between a conventional System S1, made of two monofunctional units performing the structural and TES functions, respectively, and a multifunctional material (System S2) designed to perform both functions (Fig. 5.2). System S1 could be represented by a PCM-containing unit attached to a refrigerated truck (performing the TES function) and the truck structure made in monofunctional aluminum or composite, while System S2 could be represented by a multifunctional sandwich panel containing a PCM, able to sustain the load and manage the temperature.

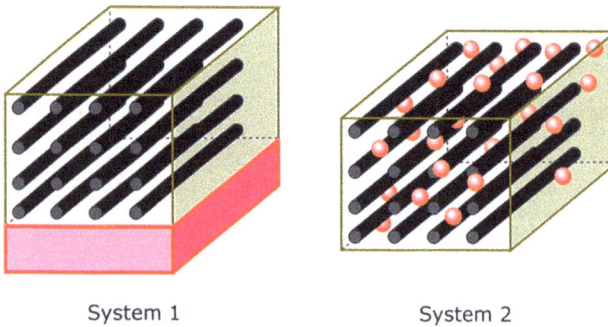

System 1   System 2

Fig. 5.2: Representative examples of conventional System S1, made of two monofunctional units performing the structural and TES functions, respectively, and a multifunctional System S2, made of a multifunctional composite designed to perform both functions.

If the two units of the conventional System 1 have masses m_s and m_{TES}, respectively, System S1 has a total mass (M) given in eq. (5.1),

$$M = m_s + m_{TES}. \tag{5.1}$$

The TES unit has a phase change enthalpy per unit mass ΔH [J/g], and the structural unit has a specific (normalized by density) elastic modulus $\bar{E}$ [GPa/(g/cm^3)]. If the TES unit does not perform any load-bearing function and the structural unit does not participate in the thermal management function, then the whole system has a total melting enthalpy of ΔH and a total specific modulus of $\bar{E}$. Hence, ΔH and $\bar{E}$ are the parameters describing the performance of the full system.

On the other hand, System 2 can be made of a structural TES material with a mass m_{mf}^*. Therefore, the total mass of System 2 (M^*) is now the mass of this structural TES material, or, as described by eq. (5.2),

$$M^* = m_{mf}^*. \tag{5.2}$$

To be worth using it, System 2, having a specific enthalpy of ΔH_{mf}^* and specific elastic modulus of $\bar{E}_{mf}^*$, should feature at least the same structural and TES performance as System 1, in terms of total absorbed and released energy [J] and total elastic modulus [GPa], or as described in eqs. (5.3) and (5.4) if a unit volume is considered:

$$\Delta H \cdot m_{TES} = \Delta H_{mf}^* \cdot m_{mf}^* \tag{5.3}$$

$$\bar{E} \cdot m_s = \bar{E}_{mf}^* \cdot m_{mf}^*. \tag{5.4}$$

These conditions can also be expressed by eq. (5.6), from which eq. (5.6) can be derived:

$$M - M^* = \frac{\bar{E}_{mf}^*}{\bar{E}} \cdot m_{mf}^* + \frac{\Delta H_{mf}^*}{\Delta H} \cdot m_{mf}^* - m_{mf}^* = \left(\frac{\bar{E}_{mf}^*}{\bar{E}} + \frac{\Delta H_{mf}^*}{\Delta H} - 1 \right) \cdot m_{mf}^* > 0 \tag{5.5}$$

$$\frac{\bar{E}_{mf}^*}{\bar{E}} + \frac{\Delta H_{mf}^*}{\Delta H} > 1. \tag{5.6}$$

We can now define a structural efficiency η_s and a TES efficiency η_{TES} via eqs. (5.7) and (5.8),

$$\eta_s = \frac{\bar{E}_{mf}^*}{\bar{E}}, \tag{5.7}$$

$$\eta_{TES} = \frac{\Delta H_{mf}^*}{\Delta H}, \tag{5.8}$$

and a multifunctional efficiency as the sum of the two, i.e., as expressed in eq. (5.9),

$$\eta_{mf} = \eta_s + \eta_{TES}. \tag{5.9}$$

Hence, using the multifunctional System 2 allows for saving mass when $\eta_{mf} > 1$. It is evident that this requirement can be met even if η_s and η_{TES} are individually lower than 1, i.e., if the multifunctional material has a specific elastic modulus and specific phase change enthalpy lower than those of the monofunctional structural unit and the monofunctional TES unit, respectively. The same analysis can be done by considering, instead of the elastic modulus, the mechanical strength. In this case, the structural efficiency would be calculated as the ratio between the specific (i.e., divided by density) strength of the multifunctional composite and the specific strength of the monofunctional structural laminate.

The presented analysis can be directly applied to the composites found in the literature and listed in Tab. 5.1, to assess if some of the prepared systems would allow an effective mass saving. The parameters η_s, η_{TES} and η_{mf} were calculated by considering as monofunctional structural units the respective composites without PCM (presenting an η_s of 1 and an η_{TES} of 0, by definition) and as the monofunctional TES unit the respective microencapsulated or shape-stabilized PCM (presenting an η_s of 0 and an η_{TES} of 1, by definition), although in a true monofunctional TES unit, the PCM would probably be somehow macro-encapsulated. The data of η_s, η_{TES}, and η_{mf} calculated with data of elastic modulus, phase change enthalpy, and density are presented in Fig. 5.3a (in this case η_s is called η_{s_E}), while the data of multifunctional efficiency considering the specific strength are presented in Fig. 5.3b (η_s is called η_{s_σ}).

In both cases, for the composites containing continuous reinforcements, a considerable mass saving is obtained for the system starch/wood/PEG, in which not only η_{mf} is greater than 1, but also η_s.itself is. This result is probably the consequence of the fact that the multifunctionality is at the phase level, as the wood laminae are both the reinforcement and the shape-stabilizing agent and PEG contributes to the mechanical properties of the laminae.

However, η_{mf} is greater than 1 also for some other systems. This means that, even though the multifunctionality of these systems is not at the level of the single constituent but at the level of the whole composite material, they allow a certain (although lower) mass saving. For the composites EP-ParCNTx-CF and PA-MCx-GF, the multifunctional efficiency increases with the PCM content. This depends on two factors: (i) in these systems, the fiber volume fraction was maintained constant, and (ii) the processing parameters were mild enough to avoid any critical modifications of most of the PCM. For the other systems, the maximum multifunctional efficiency is generally found at medium PCM concentrations, as at higher PCM concentrations not only does the PCM decrease the elastic modulus and strength *per se*, but it also contributes to decreasing the fiber fraction.

For the composites containing discontinuous fibers, for each system, the monofunctional structural component was considered as that with the same matrix type and fiber fraction, but without PCM. As mentioned previously, in these systems, the fiber content is generally lower than that of continuous fibers, and the introduction of PCMs does not influence the fiber weight fraction, even though it could sometimes slightly decrease the total fiber volume fraction because the density of the used PCMs is generally slightly lower than that of the employed polymer matrices. For the systems EP-MCx-CFS10 and PA-MCx-CFS20, the multifunctional efficiency is generally higher than 1 and increases with the PCM content. Again, this occurs because (i) the fiber volume fraction is the main parameter governing the mechanical properties and is nearly constant (high η_s), and (ii) the processing conditions preserve most of the PCM from leakage and/or degradation (high).

Although this approach allows for identifying the composition that maximizes the material's multifunctional efficiency, this analysis is valid only if it considers the most

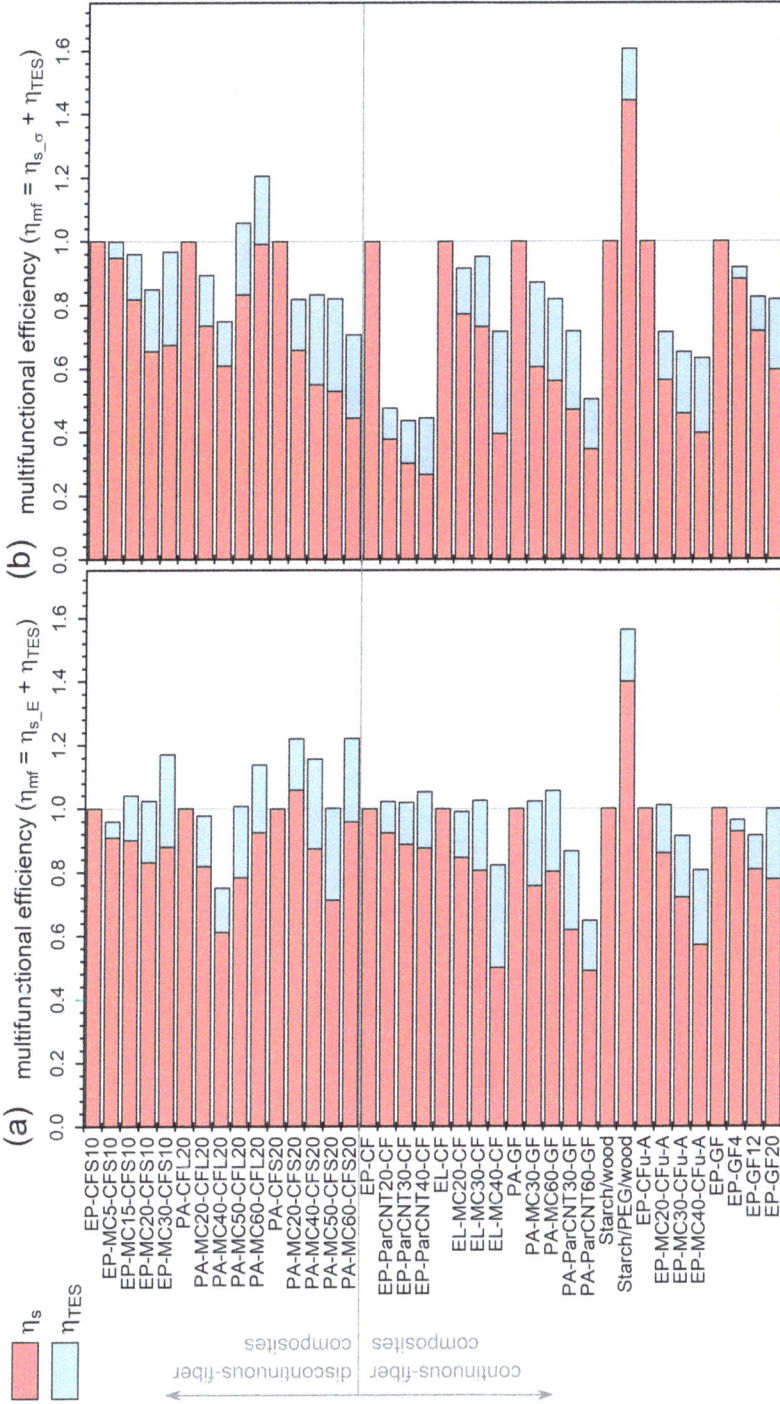

Fig. 5.3: Multifunctional efficiency of the composites presented as the sum of structural and TES efficiencies. An effective mass saving is obtained when η_{mf} is greater than 1. (a) The structural efficiency is calculated with the data of elastic modulus; (b) the structural efficiency is calculated with the data of tensile/flexural strength. Please see Table 5.1 for the meaning of the labels.

significant parameters for a specific application. In fact, for other cases, it may be more meaningful to perform a volume-saving (and not mass-saving) analysis or to maximize other mechanical properties, such as the impact energy or the fracture toughness.

To maximize η_{mf}, two routes are recommended. The first is shifting the multifunctionality from the level of the composite down to the level of the single phase. Taking as an example the case of the starch/wood/PEG laminate, it could be beneficial to find other reinforcing agents with a porous structure that can also act as shape stabilizers for the PCMs. Alternatively, it may be useful to optimize the properties of the PCM microcapsules, in terms of the shell's mechanical stiffness and adhesion with the polymer matrix. In this way, the PCM phase would contribute to the mechanical properties of the composite, i.e., would factor into the structural efficiency. This route is explored in the polydopamine-coated microcapsules described in Section 4.3.1.2 which, manifesting an increasing adhesion with the surrounding epoxy matrix and likely being stiffer, increased the mechanical properties compared to the bare epoxy or the epoxy/neat PCM composites.

The second route to maximize η_{mf} lies in the optimization of the material design and selection. The results showed that semi-structural composites reinforced with discontinuous fibers have generally higher η_{mf}, thus being a better option when maximizing the phase change enthalpy is more important than reaching very high mechanical properties. For continuous fiber composites, since the PCM tends to weaken and thicken the interlaminar region, an improved design of the stacking sequence could concentrate most of the PCM in the core layers and leave the outer layers richer in the reinforcing phase. A further extension of this concept may consider sandwich structures where all the PCM is concentrated in the core and the mechanical resistance is demanded to the outer skins, thereby shifting the multifunctionality from the material level up to the structure level. This strategy was explored in the Case Study 4 (Section 4.3.2.4), in which a multifunctional polyurethane foam containing microencapsulated PCM was used as a core in the construction of sandwich composites, in combination with monofunctional epoxy/carbon fiber skins.

6 Conclusions and future perspectives

6.1 Comprehensive overview of the research trajectory

The exploration of thermal energy storage (TES) composites presented in this monograph represents a critical investigation into the intersection of material science, energy storage, and sustainable engineering solutions. The research journey began with a comprehensive examination of the broader context of energy challenges in Chapter 1, which highlighted the pressing need for innovative energy storage technologies in response to fossil fuel depletion, climate change, and the increasing global demand for sustainable energy solutions. This foundational chapter established the critical importance of developing advanced energy storage systems capable of addressing the intermittent nature of renewable energy sources. The classification of energy storage technologies reveals a complex and diverse ecosystem. From chemical and electrical to mechanical and thermal storage systems, each approach presents unique characteristics and advantages. Thermal energy storage (TES) stands out particularly, demonstrating longer storage times and superior efficiency compared to alternative methods. Critically, TES technologies serve multiple essential purposes. They facilitate the storage and recovery of waste heat, contribute to precise temperature regulation, and provide mechanisms to prevent temperature-related damage to critical components. While thermochemical heat storage presents the highest energy density, current technological maturity remains limited. Conversely, sensible and latent heat storage systems demonstrate lower risk profiles and more immediate commercialization potential.

The subsequent chapters delved progressively deeper into the technical intricacies of TES technologies. Chapter 2 provided an in-depth analysis of latent heat thermal energy storage (LH-TES) and phase change materials (PCMs), revealing their remarkable potential for achieving high energy density and having a narrow phase change temperature range, the latter being particularly beneficial in applications where temperature stability is crucial, such as in indoor climate control or cooling electronic devices. The chapter explains the fundamental principles of latent heat storage, which involves the absorption and release of heat during phase transitions, primarily melting and crystallization. This process is highlighted as a key feature of LH-TES systems, allowing for efficient energy storage and temperature regulation. Moreover, the chapter categorizes various types of PCMs, including organic, inorganic, and eutectic mixtures, and discusses their respective properties and applications. It also addresses the challenges associated with PCMs, such as low thermal conductivity, which necessitates enhancements through additives or encapsulation to improve performance. The importance of cycle stability and compatibility with other materials is also underscored, as these factors are critical for the long-term viability of PCMs in energy storage applications.

https://doi.org/10.1515/9783111111865-006

Chapter 3 provides a comprehensive exploration of polymer matrix composites (PMCs), focusing on their definition, classification, and applications. The chapter comprehensively examines the different types of reinforcements and matrices, possible fabrication techniques, and diverse applications of these materials, emphasizing their significance across multiple industrial sectors, particularly in the aerospace and automotive industries. The insights provided in this chapter are essential for understanding how polymer-matrix composites can be effectively utilized in various engineering applications and how they can include, as an additional phase, a PCM, paving the way for advancements in composite material technology.

Building upon these foundations, Chapter 4 introduces the innovative concept of multifunctional composites, particularly focusing on their integration with thermal energy storage (TES) capabilities. The chapter begins by introducing the fundamental idea of multifunctional materials, which are designed to perform multiple functions simultaneously, thereby enhancing their utility in various applications and potentially reducing the mass and volume of the components. This is particularly relevant in the context of structural materials, where the integration of TES technology can lead to significant advancements in performance and efficiency. The chapter emphasizes the role of polymer composites in achieving multifunctionality. It discusses how these composites can be engineered to not only provide structural support but also manage thermal energy effectively. This dual functionality is crucial for applications where weight and volume savings are paramount, such as in the aerospace and automotive industries. The text highlights the potential of polymer composites to be tailored through the selection of appropriate matrices and reinforcement phases, which can optimize their properties for specific applications. Furthermore, by describing four case studies, the chapter explores the various production processes involved in creating these multifunctional composites. It outlines the challenges and opportunities associated with integrating TES into structural materials, noting that the combination can lead to innovative designs that serve as both structural components and thermal management systems. This integration is not only beneficial for performance but also contributes to sustainability by reducing the need for separate systems to handle structural integrity and thermal regulation.

However, while the integration of organic phase change materials (PCMs) into fiber-reinforced composites can enhance TES properties, it often comes at the cost of mechanical performance, particularly when high PCM weight fractions are used. This trade-off is a central theme throughout the discussion, emphasizing that the structural and TES properties of these materials are rarely synergistic.

This is the core theme of Chapter 5, which begins by discussing three primary reasons for this lack of synergy. First, the introduction of a third phase, such as a PCM, reduces the maximum fiber volume fraction, as the matrix volume fraction cannot fall below a certain threshold. Second, it points out that commercial PCMs are typically not designed to serve as fillers that enhance the mechanical properties of polymer matrices. Consequently, their thermal, mechanical, and surface properties are not optimized for reinforcing roles. Lastly, the reinforcement materials seldom act as shape-stabilizing agents for the PCM, which could otherwise maintain mechanical properties when the PCM is introduced. Despite these challenges, combining structural and TES functions can still offer significant advantages in terms of mass and volume savings compared to using separate monofunctional units. To quantify these benefits, the development of objective selection and design criteria is proposed that take into account the competing requirements of structural TES composites. The chapter introduces a multifunctionality parameter, called multifunctional efficiency, aimed at minimizing the mass of components that fulfill both thermal and structural roles, thereby facilitating the evaluation and ranking of various structural TES composites described in the open scientific literature. The proposed multifunctionality parameter represents a critical step towards objective material selection and design criteria, offering a framework for a quantitative evaluation of structural TES composites. This approach is essential for establishing design principles and material selection guidelines that maximize the potential of multifunctional materials in practical applications. The presented analysis reveals that, although the introduction of a PCM rarely increases the mechanical properties of the composite (structural efficiency lower than 1), there still can be an advantage in the total mass saving (multifunctional efficiency higher than 1), especially when the fiber volume fraction is kept as constant as possible or when the single phases are themselves multifunctional, i.e., when the reinforcement also acts as a shape-stabilizing agent or when the PCM also contributes to the mechanical performance.

6.2 Future research directions

The field of multifunctional structural thermal energy storage (TES) composites stands at a critical juncture of materials science and sustainable engineering, presenting both significant challenges and transformative potential. The future development of these materials will require a sophisticated, multidimensional approach that transcends traditional material design paradigms.

The most promising avenue for advancement lies in fundamentally reimagining the relationship between composite constituents. Rather than treating structural and thermal energy storage capabilities as competing functions, researchers must pursue strategies that enable genuine synergy at the phase level. This approach demands a radical rethinking of material constituent properties, focusing on creating reinforcing agents with inherently multifunctional characteristics. Specifically, future research should prioritize developing reinforcing phases with multifunctional capabilities beyond structural support. Porous structures that simultaneously act as shape stabilizers for PCMs represent a particularly compelling direction. Such innovations could overcome the current limitations where PCM introduction typically compromises mechanical performance. The goal is to engineer composite phases that contribute simultaneously to mechanical strength and thermal energy storage, rather than creating materials where these properties coexist in separate phases. This approach has been key in one of the case studies presented in Chapter 4, in which the thin wood laminae also acted as shape-stabilizing agents for PEG.

The optimization of PCM microcapsules emerges as another critical research frontier. By enhancing the mechanical properties of capsule shells and improving their matrix adhesion, researchers can potentially transform PCMs from passive thermal storage elements to active contributors to composite mechanical performance. This approach requires sophisticated materials engineering, potentially leveraging advanced coating technologies and precision manufacturing techniques.

Material design strategies must also evolve to accommodate more nuanced performance requirements. For semi-structural composites with discontinuous fibers, the focus should shift towards maximizing phase change enthalpy while maintaining acceptable mechanical characteristics. Continuous fiber composites, conversely, will benefit from innovative stacking sequences that strategically concentrate PCMs in the core layers and optimize reinforcement distribution. Indeed, the concept of structure-level multifunctionality introduces another exciting research direction, which presents itself as an alternative to designing multifunctional phases. Sandwich structure designs that segregate thermal storage and mechanical load-bearing functions across different layers offer a promising approach. By concentrating PCMs in core layers and utilizing high-performance materials for external skins, researchers can create composites that maximize both thermal management and structural integrity.

Critically, these advancements must be underpinned by robust quantitative methodologies for assessing multifunctionality. The development of comprehensive selection and design criteria becomes paramount. Metrics like multifunctional efficiency provide a framework for objective material evaluation, enabling researchers to systematically compare and optimize composite designs across various applications.

The ultimate objective transcends mere technical achievement. These multifunctional structural TES composites represent a pivotal technology in addressing global sustainability challenges. By enabling more efficient energy storage, precise temperature regulation, and material performance optimization, this research domain contributes directly to mitigating climate change and supporting the transition to renewable energy systems. The path forward requires interdisciplinary collaboration, combining expertise from materials science, thermal engineering, computational modeling, and sustainable design. Success will demand not only scientific innovation but also a holistic understanding of how these advanced materials can be integrated into broader technological ecosystems.

As research progresses, multifunctional composites in general, and multifunctional structural TES composites in particular, will likely emerge as a cornerstone technology, offering unprecedented capabilities in energy management, material efficiency, and sustainable engineering solutions. The journey from current limitations to future potential represents a testament to human ingenuity and our capacity to reimagine material performance at the most fundamental levels.

References

[1] S. Kalaiselvam and R. Parameshwaran, Eds. *Thermal Energy Storage Technologies for Sustainability: Systems Design, Assessment, and Applications*, 2014, Academic Press, Elsevier, London, UK.

[2] I. Dincer and M. A. Ezan, *Heat Storage: A Unique Solution for Energy Systems*, Springer International, 2018, Cham, Switzerland.

[3] I. Dincer and M. A. Rosen, *Thermal Energy Storage. Systems and Applications*, 2 ed. Chichester, West Sussex, PO19 8SQ United Kingdom: John Wiley and Sons, Ltd, 2011.

[4] L. F. Cabeza and L. F. Cabeza, Ed. *Advances in Thermal Energy Storage Systems: Methods and Applications*, (Woodhead Publishing Series in Energy). 80 High Street, Sawston, Cambridge, CB22 3HJ UK:: Woodhead Publishing, 2014.

[5] G. Fredi, A. Dorigato, L. Fambri and A. Pegoretti, Multifunctional structural composites for thermal energy storage, *Multifunctional Materials*, vol. 3, pp. 042001, 2020.

[6] J. Zhang, -H.-H. Zhang, Y.-L. He and W.-Q. Tao, A comprehensive review on advances and applications of industrial heat pumps based on the practices in China, *Applied Energy*, vol. 178, pp. 800–825, 2016.

[7] N. Sheng, C. Zhu, H. Sakai, T. Akiyama and T. Nomura, Synthesis of Al-25 wt% Si@Al2O3@Cu microcapsules as phase change materials for high temperature thermal energy storage, *Solar Energy Materials & Solar Cells*, vol. 191, pp. 141–147, 2019.

[8] M. Ostry and P. Charvat, Materials for advanced heat storage in buildings, in *11th International Conference on Modern Building Materials, Structures and Techniques (MBMST 2013)*, Vilnius, Lithuania, vol. 57, pp. 837–843, 2013. Procedia Engineering.

[9] C. Cherif, N. H. A. Tran, M. Kirsten, H. Bruenig and R. Vogel, Environmentally friendly and highly productive bi-component melt spinning of thermoregulated smart polymer fibres with high latent heat capacity, *Express Polymer Letters*, vol. 12, no. 3, pp. 203–214, 2018.

[10] R. Kandasamy, X.-Q. Wang and A. S. Mujumdar, Application of phase change materials in thermal management of electronics, *Applied Thermal Engineering*, vol. 27, no. 17–18, pp. 2822–2832, 2007.

[11] H. M. Ali, *et al.* Advances in thermal energy storage: Fundamentals and applications, *Progress in Energy and Combustion Science*, vol. 100, pp. 101109, 2024.

[12] A. Safari, R. Saidur, F. A. Sulaiman, Y. Xu and J. Dong, A review on supercooling of Phase Change Materials in thermal energy storage systems, *Renewable and Sustainable Energy Reviews*, vol. 70, pp. 905–919, 2017.

[13] H. Jarimi, D. Aydin, Z. Yanan, G. Ozankaya, X. Chen and S. Riffat, Review on the recent progress of thermochemical materials and processes for solar thermal energy storage and industrial waste heat recovery, *International Journal of Low-Carbon Technologies*, vol. 14, no. 1, pp. 44–69, 2019.

[14] S. M. Hasnain, Review on sustainable thermal energy storage technologies, Part I: heat storage materials and techniques, *Energy Conversion and Management*, vol. 39, no. 11, pp. 1127–1138, 1998.

[15] D. Aydin, S. P. Casey and S. Riffat, The latest advancements on thermochemical heat storage systems, *Renewable and Sustainable Energy Reviews*, vol. 41, pp. 356–367, 2015.

[16] Z. Ye, H. Liu, W. Wang, H. Liu, J. Lv and F. Yang, Reaction/sorption kinetics of salt hydrates for thermal energy storage, *Journal of Energy Storage*, vol. 56, pp. 106122, 2022.

[17] S. Vasta, *et al.* Adsorption Heat Storage: State-of-the-Art and Future Perspectives, *Nanomaterials (Basel)*, vol. 8, no. 7, pp. 106122, Jul 12 2018.

[18] V. Kulish, N. Aslfattahi, M. Schmirler and P. Slama, New library of phase-change materials with their selection by the Renyi entropy method, *Scientific Reports*, vol. 13, no. 1, pp. 10446, Jun 27 2023.

[19] K. Pielichowska and K. Pielichowski, Phase change materials for thermal energy storage, *Progress in Materials Science*, vol. 65, pp. 67–123, 2014.

https://doi.org/10.1515/9783111111865-007

[20] R. K. Sharma, P. Ganesan, V. V. Tyagi, H. S. C. Metselaar and S. C. Sandaran, Developments in organic solid-liquid phase change materials and their applications in thermal energy storage, (in English), *Energy Conversion and Management*, Article vol. 95, pp. 193–228, May 2015.

[21] S. Kahwaji, M. B. Johnson, A. C. Kheirabadi, D. Groulx and M. A. White, A comprehensive study of properties of paraffin phase change materials for solar thermal energy storage and thermal management applications, *Energy*, vol. 162, pp. 1169–1182, 2018.

[22] G. A. Lane, *Solar Heat Storage Latent Heat Materials, Volume 1: Background and Scientific Principles*, CRC Press, Boca Raton, FL, USA, 1983.

[23] N. Sarier and E. Onder, Organic phase change materials and their textile applications: An overview, *Thermochimica Acta*, vol. 540, pp. 7–60, 2012.

[24] S. Sundararajan, A. B. Samui and P. S. Kulkarni, Versatility of polyethylene glycol (PEG) in designing solid–solid phase change materials (PCMs) for thermal management and their application to innovative technologies, *Journal of Materials Chemistry A*, vol. 5, no. 35, pp. 18379–18396, 2017.

[25] K. Król, B. Macherzyńska and K. Pielichowska, Acrylic bone cements modified with poly(ethylene glycol)-based biocompatible phase-change materials, *Journal of Applied Polymer Science*, vol. 133, no. 36, pp. 43898, 2016.

[26] A. S. Fleischer, *Thermal Energy Storage Using Phase Change Materials – Fundamentals and Applications*, Minneapolis, MN, USA: Springer Briefs in Applied Science and Technology, Thermal Engineering and Applied Science 2015.

[27] I. Dincer, H. S. Hamut and N. Javani, *Thermal Management of Electric Vehicle Battery Systems*, John Wiley & Sons, Chichester, UK, 2017.

[28] B. Zalba, J. M. Marin, L. F. Cabeza and H. Mehling, Review on thermal energy storage with phase change: materials, heat transfer analysis and applications, *Applied Thermal Engineering*, vol. 23, pp. 251–283, 2003.

[29] G. B. Hamad, Z. Younsi, H. Naji and F. Salaün, A Comprehensive Review of Microencapsulated Phase Change Materials Synthesis for Low-Temperature Energy Storage Applications, *Applied Sciences*, vol. 11, no. 24, pp. 11900, 2021.

[30] B. Zalba, J. Marin, L. F. Cabeza and H. Mehling, Review on thermal energy storage with phase change: materials, heat transfer analysis and applications, *Applied Thermal Engineering*, vol. 23, pp. 251–283, 2003.

[31] A. S. Fleischer, *Thermal Energy Storage Using Phase Change Materials: Fundamentals and Applications*, Springer International Publishing AG Switzerland, Cham, Switzerland, 2015.

[32] A. Jamekhorshid, S. M. Sadrameli and M. Farid, A review of microencapsulation methods of phase change materials (PCMs) as a thermal energy storage (TES) medium, *Renewable and Sustainable Energy Reviews*, vol. 31, pp. 531–542, 2014.

[33] Y. Konuklu, M. Ostry, H. O. Paksoy and P. Charvat, Review on using microencapsulated phase change materials (PCM) in building applications, *Energy and Buildings*, vol. 106, pp. 134–155, 2015.

[34] A. Palacios, M. E. Navarro-Rivero, B. Zou, Z. Jiang, M. T. Harrison and Y. Ding, A perspective on Phase Change Material encapsulation: Guidance for encapsulation design methodology from low to high-temperature thermal energy storage applications, *Journal of Energy Storage*, vol. 72, pp. 108597, 2023.

[35] S. Shoeibi, *et al.* Recent advancements in applications of encapsulated phase change materials for solar energy systems: A state of the art review, *Journal of Energy Storage*, vol. 94, pp. 112401, 2024.

[36] C. Y. Zhao and G. H. Zhang, Review on microencapsulated phase change materials (MEPCMs): Fabrication, characterization and applications, *Renewable and Sustainable Energy Reviews*, vol. 15, no. 8, pp. 3813–3832, 2011.

[37] V. V. Tyagi, S. C. Kaushik, S. K. Tyagi and T. Akiyama, Development of phase change materials based microencapsulated technology for buildings: A review, *Renewable and Sustainable Energy Reviews*, vol. 15, no. 2, pp. 1373–1391, 2011.

[38] L. Y. Wang, P. S. Tsai and Y. M. Yang, Preparation of silica microspheres encapsulating phase-change material by sol-gel method in O/W emulsion, *Journal of Microencapsulation*, vol. 23, no. 1, pp. 3–14. Feb 2006.

[39] R. Ciriminna, M. Sciortino, G. Alonzo, A. de schrijver and M. Pagliaro, From Molecules to Systems: Sol-Gel Microencapsulation in Silica-Based Materials, *Chemical Reviews*, vol. 111, no. 2, pp. 765–789, 2011.

[40] Y. Lin, C. Zhu and G. Fang, Synthesis and properties of microencapsulated stearic acid/silica composites with graphene oxide for improving thermal conductivity as novel solar thermal storage materials, *Solar Energy Materials and Solar Cells*, vol. 189, pp. 197–205, 2019.

[41] Z. Chen, L. Cao, G. Fang and F. Shan, Synthesis and Characterization of Microencapsulated Paraffin Microcapsules as Shape-Stabilized Thermal Energy Storage Materials, *Nanoscale and Microscale Thermophysical Engineering*, vol. 17, no. 2, pp. 112–123, 2013.

[42] F. Tang, L. Liu, G. Alva, Y. Jia and G. Fang, Synthesis and properties of microencapsulated octadecane with silica shell as shape–stabilized thermal energy storage materials, *Solar Energy Materials and Solar Cells*, vol. 160, pp. 1–6, 2017.

[43] S. Freitas, H. P. Merkle and B. Gander, Microencapsulation by solvent extraction/evaporation: reviewing the state of the art of microsphere preparation process technology, *Journal of Controlled Release*, vol. 102, no. 2, pp. 313–332, Feb 2 2005.

[44] M. M. Umair, Y. Zhang, K. Iqbal, S. Zhang and B. Tang, Novel strategies and supporting materials applied to shape-stabilize organic phase change materials for thermal energy storage–A review, *Applied Energy*, vol. 235, pp. 846–873, 2019.

[45] Q. Zhang and J. Liu, Anisotropic thermal conductivity and photodriven phase change composite based on RT100 infiltrated carbon nanotube array, *Solar Energy Materials and Solar Cells*, vol. 190, pp. 1–5, 2019.

[46] Y. Xia, *et al.* Graphene-oxide-induced lamellar structures used to fabricate novel composite solid-solid phase change materials for thermal energy storage, *Chemical Engineering Journal*, vol. 362, pp. 909–920, 2019.

[47] P. Zhang, Y. Hu, L. Song, J. Ni, W. Xing and J. Wang, Effect of expanded graphite on properties of high-density polyethylene/paraffin composite with intumescent flame retardant as a shape-stabilized phase change material, *Solar Energy Materials and Solar Cells*, vol. 94, no. 2, pp. 360–365, 2010.

[48] A. Sarı and A. Karaipekli, Thermal conductivity and latent heat thermal energy storage characteristics of paraffin/expanded graphite composite as phase change material, *Applied Thermal Engineering*, vol. 27, no. 8–9, pp. 1271–1277, 2007.

[49] K. Biswas, J. Lu, P. Soroushian and S. Shrestha, Combined experimental and numerical evaluation of a prototype nano-PCM enhanced wallboard, *Applied Energy*, vol. 131, pp. 517–529, 2014.

[50] N. Xie, *et al.* Salt hydrate/expanded vermiculite composite as a form-stable phase change material for building energy storage, *Solar Energy Materials and Solar Cells*, vol. 189, pp. 33–42, 2019.

[51] X. Fang, *et al.* Thermal energy storage performance of paraffin-based composite phase change materials filled with hexagonal boron nitride nanosheets, *Energy Conversion and Management*, vol. 80, pp. 103–109, Apr 2014.

[52] W. Yuan, X. Yang, G. Zhang and X. Li, A thermal conductive composite phase change material with enhanced volume resistivity by introducing silicon carbide for battery thermal management, *Applied Thermal Engineering*, vol. 144, pp. 551–557, 2018.

[53] X. Huang, Y. X. Lin, G. Alva and G. Y. Fang, Thermal properties and thermal conductivity enhancement of composite phase change materials using myristyl alcohol/metal foam for solar thermal storage, (in English), *Solar Energy Materials and Solar Cells*, Article vol. 170, pp. 68–76, Oct 2017.

[54]   H. Luo, F. Yu, H. Wu, C. Wang and M. Wang, High-Performance Heat Storage Phase Change Composite by Highly Aligned N-Doped Mesoporous Carbon for Efficient Battery Thermal Management, *ACS Applied Energy Materials*, vol. 6, no. 2, pp. 950–959, 2023.

[55]   K. Resch-Fauster and M. Feuchter, Thermo-physical characteristics, mechanical performance and long-term stability of high temperature latent heat storages based on paraffin-polymer compounds, *Thermochimica Acta*, vol. 663, pp. 34–45, 2018.

[56]   A. Kylili and P. A. Fokaides, Life Cycle Assessment (LCA) of Phase Change Materials (PCMs) for building applications: A review, *Journal of Building Engineering*, vol. 6, pp. 133–143, 2016.

[57]   K. Struhala and M. Ostrý, Life-Cycle Assessment of phase-change materials in buildings: A review, *Journal of Cleaner Production*, vol. 336, pp. 130359, 2022.

[58]   Z. A. Qureshi, H. M. Ali and S. Khushnood, Recent advances on thermal conductivity enhancement of phase change materials for energy storage system: A review, *International Journal of Heat and Mass Transfer*, vol. 127, pp. 838–856, 2018.

[59]   Y. Lin, Y. Jia, G. Alva and G. Fang, Review on thermal conductivity enhancement, thermal properties and applications of phase change materials in thermal energy storage, *Renewable and Sustainable Energy Reviews*, vol. 82, pp. 2730–2742, 2018.

[60]   L. Liu, D. Su, Y. Tang and G. Fang, Thermal conductivity enhancement of phase change materials for thermal energy storage: A review, *Renewable and Sustainable Energy Reviews*, vol. 62, pp. 305–317, 2016.

[61]   B. Eanest Jebasingh and A. Valan Arasu, A comprehensive review on latent heat and thermal conductivity of nanoparticle dispersed phase change material for low-temperature applications, *Energy Storage Materials*, vol. 24, pp. 52–74, 2020.

[62]   L. Fan and J. M. Khodadadi, Thermal conductivity enhancement of phase change materials for thermal energy storage: A review, *Renewable and Sustainable Energy Reviews*, vol. 15, no. 1, pp. 24–46, 2011.

[63]   X. Cao, X. Dai and J. Liu, Building energy-consumption status worldwide and the state-of-the-art technologies for zero-energy buildings during the past decade, *Energy and Buildings*, vol. 128, pp. 198–213, 2016.

[64]   F. Valentini, G. Fredi and A. Dorigato, Thermal Energy Storage (TES) for Sustainable Buildings: Addressing the Current Energetic Situation in the EU with TES-Enhanced Buildings, in *Natural Energy, Lighting, and Ventilation in Sustainable Buildings*, M. Nazari-Heris, Ed. Cham: Springer, pp. 191–215, 2024.

[65]   R. Aridi and A. Yehya, Review on the sustainability of phase-change materials used in buildings, *Energy Conversion and Management: X*, vol. 15, pp. 100237, 2022.

[66]   M. Ghamari, *et al.* Advancing sustainable building through passive cooling with phase change materials, a comprehensive literature review, *Energy and Buildings*, vol. 312, pp. 114164, 2024.

[67]   B. M. Tripathi and S. K. Shukla, A comprehensive review of the thermal performance in energy efficient building envelope incorporated with phase change materials, *Journal of Energy Storage*, vol. 79, pp. 110128, 2024.

[68]   V. J. Reddy, M. F. Ghazali and S. Kumarasamy, Advancements in phase change materials for energy-efficient building construction: A comprehensive review, *Journal of Energy Storage*, vol. 81, pp. 110494, 2024.

[69]   M. Kenisarin and K. Mahkamov, Passive thermal control in residential buildings using phase change materials, *Renewable and Sustainable Energy Reviews*, vol. 55, pp. 371–398, 2016.

[70]   D. Feldman and D. Banu, DSC analysis for the evaluation of an energy storing wallboard, *Thermochimica Acta*, vol. 272, pp. 243–251, 1996.

[71]   S. Scalat, D. Banu, D. Hawes, J. Paris, F. Haghighata and D. Feldman, Full scale thermal testing of latent heat storage in wallboard, *Solar Energy Materials and Solar Cells*, vol. 44, pp. 49–61, 1996.

[72]   Q. Wang and C. Y. Zhao, Parametric investigations of using a PCM curtain for energy efficient buildings, *Energy and Buildings*, vol. 94, pp. 33–42, 2015.

[73] H. Akeiber, *et al.* A review on phase change material (PCM) for sustainable passive cooling in building envelopes, *Renewable and Sustainable Energy Reviews*, vol. 60, pp. 1470–1497, 2016.

[74] X. Bao, *et al.* Development of high performance PCM cement composites for passive solar buildings, *Energy and Buildings*, vol. 194, pp. 33–45, 2019.

[75] K. Lin, Y. Zhang, X. Xu, H. Di, R. Yang and P. Qin, Experimental study of under-floor electric heating system with shape-stabilized PCM plates, *Energy and Buildings*, vol. 37, no. 3, pp. 215–220, 2005.

[76] K. Yang, *et al.* Review: incorporation of organic PCMs into textiles, *Journal of Materials Science*, vol. 57, no. 2, pp. 798–847, 2022.

[77] M. T. Hossain, M. A. Shahid, M. Y. Ali, S. Saha, M. S. I. Jamal and A. Habib, Fabrications, Classifications, and Environmental Impact of PCM-Incorporated Textiles: Current State and Future Outlook, *ACS Omega*, vol. 8, no. 48, pp. 45164–45176, Dec 5 2023.

[78] M. M. Pritom, *et al.* Phase change materials in textiles: synthesis, properties, types and applications – a critical review, *Textile Research Journal*, pp. 2763–2779, 2024.

[79] S. Mondal, Phase change materials for smart textiles – An overview, *Applied Thermal Engineering*, vol. 28, no. 11–12, pp. 1536–1550, 2008.

[80] K. Iqbal, *et al.* Phase change materials, their synthesis and application in textiles – a review, *The Journal of The Textile Institute*, vol. 110, no. 4, pp. 625–638, 2019.

[81] K. Iqbal and D. Sun, Development of thermal stable multifilament yarn containing micro-encapsulated phase change materials, *Fibers and Polymers*, vol. 16, no. 5, pp. 1156–1162, 2015.

[82] M. H. Hartmann, J. B. Worley and M. North, *Temperature Regulating Cellulose Fibers and Applications Thereof*, USA, US Patent, US20070026228A1, Outlast Technologies, Inc. 2007.

[83] Outlast®. (2018, 20 Sept 2019). *Applying Outlast® technology to your line*. Available: http://www.out last.com/en/applications/

[84] A. Mitra, R. Kumar, D. K. Singh and Z. Said, Advances in the improvement of thermal-conductivity of phase change material-based lithium-ion battery thermal management systems: An updated review, *Journal of Energy Storage*, vol. 53, pp. 105195, 2022.

[85] S. C. Fok, W. Shen and F. L. Tan, Cooling of portable hand-held electronic devices using phase change materials in finned heat sinks, *International Journal of Thermal Sciences*, vol. 49, no. 1, pp. 109–117, 2010.

[86] S. K. Sahoo, M. K. Das and P. Rath, Application of TCE-PCM based heat sinks for cooling of electronic components: A review, *Renewable and Sustainable Energy Reviews*, vol. 59, pp. 550–582, 2016.

[87] F. L. Tan and C. P. Tso, Cooling of mobile electronic devices using phase change materials, *Applied Thermal Engineering*, vol. 24, no. 2–3, pp. 159–169, 2004.

[88] Y. Tomizawa, K. Sasaki, A. Kuroda, R. Takeda and Y. Kaito, Experimental and numerical study on phase change material (PCM) for thermal management of mobile devices, *Applied Thermal Engineering*, vol. 98, pp. 320–329, 2016.

[89] L. Ianniciello, P. H. Biwolé and P. Achard, Electric vehicles batteries thermal management systems employing phase change materials, *Journal of Power Sources*, vol. 378, pp. 383–403, 2018.

[90] S. Al Hallaj and J. R. Selman, A Novel Thermal Management System for Electric Vehicle Batteries Using Phase-Change Material, *Journal of The Electrochemical Society*, vol. 147, no. 9, pp. 3231–3236, 2000.

[91] P. Goli, S. Legedza, A. Dhar, R. Salgado, J. Renteria and A. A. Balandin, Graphene-enhanced hybrid phase change materials for thermal management of Li-ion batteries, *Journal of Power Sources*, vol. 248, pp. 37–43, 2014.

[92] D. Zou, *et al.* Preparation of a novel composite phase change material (PCM) and its locally enhanced heat transfer for power battery module, *Energy Conversion and Management*, vol. 180, pp. 1196–1202, 2019.

[93] P. Qin, M. Liao, D. Zhang, Y. Liu, J. Sun and Q. Wang, Experimental and numerical study on a novel hybrid battery thermal management system integrated forced-air convection and phase change material, *Energy Conversion and Management*, vol. 195, pp. 1371–1381, 2019.

[94] Y. Lv, Y. Zou and L. Yang, Feasibility study for thermal protection by microencapsulated phase change micro/nanoparticles during cryosurgery, *Chemical Engineering Science*, vol. 66, no. 17, pp. 3941–3953, 2011.

[95] S. Singh, K. K. Gaikwad and Y. S. Lee, Phase change materials for advanced cooling packaging, *Environmental Chemistry Letters*, vol. 16, no. 3, pp. 845–859, 2018.

[96] E. Alehosseini and S. M. Jafari, Micro/nano-encapsulated phase change materials (PCMs) as emerging materials for the food industry, *Trends in Food Science & Technology*, vol. 91, pp. 116–128, 2019.

[97] J. H. Johnston, J. E. Grindrod, M. Dodds and K. Schimitschek, Composite nano-structured calcium silicate phase change materials for thermal buffering in food packaging, *Current Applied Physics*, vol. 8, no. 3–4, pp. 508–511, 2008.

[98] M. Ünal, Y. Konuklu and H. Paksoy, Thermal buffering effect of a packaging design with microencapsulated phase change material, *International Journal of Energy Research*, vol. 43, no. 9, pp. 4495–4505, 2019.

[99] Y. Cai, *et al.* Preparation and flammability of high density polyethylene/paraffin/organophilic montmorillonite hybrids as a form stable phase change material, *Energy Conversion and Management*, vol. 48, no. 2, pp. 462–469, 2007.

[100] G. Song, S. Ma, G. Tang, Z. Yin and X. Wang, Preparation and characterization of flame retardant form-stable phase change materials composed by EPDM, paraffin and nano magnesium hydroxide, *Energy*, vol. 35, no. 5, pp. 2179–2183, 2010.

[101] L. Li, G. Wang and C. Guo, Influence of intumescent flame retardant on thermal and flame retardancy of eutectic mixed paraffin/polypropylene form-stable phase change materials, *Applied Energy*, vol. 162, pp. 428–434, 2016.

[102] P. Zhang, L. Song, H. Lu, J. Wang and Y. Hu, The influence of expanded graphite on thermal properties for paraffin/high density polyethylene/chlorinated paraffin/antimony trioxide as a flame retardant phase change material, *Energy Conversion and Management*, vol. 51, no. 12, pp. 2733–2737, 2010.

[103] B. D. Agarwal, L. J. Broutman and K. Chandrashekhara, *Analysis and Performance of Fiber Composites*, 4 ed. John Wiley and Sons, Inc, Chichester, UK, 2018.

[104] F. C. Campbell, *Structural Composite Materials*, Materials Park, Ohio, US: ASM International, 2010.

[105] R. Petrucci and L. Torre, Filled Polymer Composites, in *Modification of Polymer Properties*, C. F. Jasso-Gastinel and J. M. Kenny, Eds. Chadds Ford, PA, USA: Elsevier, 2017, pp. 23–46.

[106] R. Wang, S. Zheng and Y. Zheng, *Polymer Matrix Composites and Technology*, (Woodhead Publishing Series in Composites Science and Engineering) Beijing, China: Woodhead Publishing – Science Press Limited, 2011.

[107] R. Talreja and J. Varna, *Modeling Damage, Fatigue and Failure of Composite Materials*, Sawston, Cambridge, UK: Woodhead Publishing Series in Composites Science and Engineering, 2015.

[108] P. K. Mallick and P. K. Mallick, Ed. *Fiber Reinforced Composites. Materials, Manufacturing, and Design*, Boca Raton, FL, US: Taylor & Francis Group, LLC, 2007.

[109] A. C. Long, *Design and Manufacture of Textile Composites*, Boca Raton, FL, US: Woodhead Publishing Limited and CRC Press LLC, 2006.

[110] R. N. Rothon, *Particulate-Filled Polymer Composites*, 2 ed. Shawbury, Shrewsbury, Shropshire, SY4 4NR, UK: Rapra Technology Limited, 2003.

[111] A. V. Shenoy, *Rheology of Filled Polymer Systems*, P.O. Box 17, 3300 AA Dordrecht The Netherlands: Kluwer Academic Publishers, 1999.

[112]   M. Biron, M. Biron and O. Marichal, Eds. *Thermoplastics and Thermoplastic Composites*, Oxford, UK: Elsevier, Ltd, 2013.

[113]   L. H. Sperling, *Introduction to Physical Polymer Science*, Hoboken, New Jersey, US: Wiley, 2006.

[114]   R. Q. Shen, L. C. Hatanaka, L. Ahmed, R. J. Agnew, M. S. Mannan and Q. S. Wang, Cone calorimeter analysis of flame retardant poly (methyl methacrylate)-silica nanocomposites, (in English), *Journal of Thermal Analysis and Calorimetry*, Article vol. 128, no. 3, pp. 1443–1451, Jun 2017.

[115]   N. Burger, A. Laachachi, M. Ferriol, M. Lutz, V. Toniazzo and D. Ruch, Review of thermal conductivity in composites: Mechanisms, parameters and theory, *Progress in Polymer Science*, vol. 61, pp. 1–28, 2016.

[116]   S. K. Bhudolia, P. Perrotey and S. C. Joshi, Enhanced vibration damping and dynamic mechanical characteristics of composites with novel pseudo-thermoset matrix system, *Composite Structures*, vol. 179, pp. 502–513, 2017.

[117]   S. K. Bhudolia, P. Perrotey and S. C. Joshi, Mode I fracture toughness and fractographic investigation of carbon fibre composites with liquid Methylmethacrylate thermoplastic matrix, *Composites Part B: Engineering*, vol. 134, pp. 246–253, 2018.

[118]   A. K. Kulshreshtha and C. Vasile, *Handbook of Polymer Blends and Composites*, Shawbury, Shrewsbury, Shropshire, SY4 4NR, UK: Smithers Rapra Technology, 2002.

[119]   N. Wiegand and E. Mäder, Commingled Yarn Spinning for Thermoplastic/Glass Fiber Composites, *Fibers*, vol. 5, no. 3, pp. 26, 2017.

[120]   P.-C. Hsiao, C.-M. Lin, C.-T. Lu, W. Yin, Y.-T. Huang and J.-H. Lin, Manufacture and evaluations of stainless steel/rayon/bamboo charcoal functional composite knits, *Textile Research Journal*, vol. In Press, pp. 3893–3899, 2019.

[121]   J.-V. Risicato, *et al.* A complex shaped reinforced thermoplastic composite part made of commingled yarns with integrated sensor, *Applied Composite Materials*, vol. 22, no. 1, pp. 81–98, 2014.

[122]   M. M. Shokrieh and H. Moshrefzadeh-Sani, On the constant parameters of Halpin-Tsai equation, *Polymer*, vol. 106, pp. 14–20, 2016.

[123]   R. M. Guedes, *Creep and Fatigue in Polymer Matrix Composites*, Woodhead Publishing in Materials Abington, Cambridge CB21 6AH, UK: Woodhead Publishing, 2011.

[124]   L. A. Carlsson, D. F. Adams and R. B. Pipes, *Experimental Characterization of Advanced Composite Materials*, 4 ed. Boca Raton, FL: CRC Press, Taylor & Francis Group, 2014.

[125]   B. D. Agarwal, L. J. Broutman and K. Chandrashekhara, *Analysis and Performance of Fiber Composites*, 3 ed. Daryaganj, New Delhi: John Wiley & Sons, Inc, 2006.

[126]   V. Tuninetti, *et al. Fiber-Reinforced Composite Materials: Characterization and Computational Predictions of Mechanical Performance* (Synthesis Lectures on Mechanical Engineering), Cham, Switzerland: Springer, 2023.

[127]   S.-J. Park and M.-K. Seo, *Interface Science and Composites*, (no. INTERFACE SCIENCE AND TECHNOLOGY– VOLUME 18) Amsterdam, The Netherlands: Academic Press, Elsevier Ltd, 2011.

[128]   S. Arora, R. Chitkara, A. S. Dhangar, D. Dubey, R. Kumar and A. Gupta, A review of fatigue behavior of FRP composites, *Materials Today: Proceedings*, vol. 64, pp. 1272–1275, 2022.

[129]   C. H. Kumar, A. Bongale and C. S. Venkatesha, Factors Affecting the Fatigue Behavior of Fiber-Reinforced Polymer Matrix Composites, *Journal of The Institution of Engineers (India): Series C*, vol. 104, no. 3, pp. 647–659, 2023.

[130]   K. Friedrich, Routes for Achieving Multifunctionality in Reinforced Polymers and Composite Structures, in *Multifunctionality of Polymer Composites: Challenges and New Solutions*, K. Friedrich and U. Breuer, Eds. Waltham, MA, US: Elsevier, 2015, pp. 3–41.

[131]   L. E. Asp and E. S. Greenhalgh, Structural power composites, *Composites Science and Technology*, vol. 101, pp. 41–61, 2014.

[132]   R. F. Gibson, A review of recent research on mechanics of multifunctional composite materials and structures, *Composite Structures*, vol. 92, no. 12, pp. 2793–2810, 2010.

[133] R. Gray, *et al.* Carbon fibre based electrodes for structural batteries, *Journal of Materials Chemistry A*, vol. 12, no. 38, pp. 25580–25599, 2024.

[134] G. Fredi, *et al.* Graphitic microstructure and performance of carbon fibre Li-ion structural battery electrodes, *Multifunctional Materials*, vol. 1, no. 1, pp. 015003, 2018.

[135] K. Salonitis, J. Pandremenos, J. Paralikas and G. Chryssolouris, Multifunctional materials: Engineering applications and processing challenges, *International Journal of Advanced Manufacturing Technology*, vol. 49, no. 5–8, pp. 803–826, 2010.

[136] Z. Tian, Y. Li, J. Zheng and S. Wang, A state-of-the-art on self-sensing concrete: Materials, fabrication and properties, *Composites Part B: Engineering*, vol. 177, pp. 107437, 2019.

[137] I. Papanikolaou, N. Arena and A. Al-Tabbaa, Graphene nanoplatelet reinforced concrete for self-sensing structures – A lifecycle assessment perspective, *Journal of Cleaner Production*, vol. 240, pp. 118202, 2019.

[138] H. Qin, S. Ding, A. Ashour, Q. Zheng and B. Han, Revolutionizing infrastructure: The evolving landscape of electricity-based multifunctional concrete from concept to practice, *Progress in Materials Science*, vol. 145, pp. 101310, 2024.

[139] R. Natesan and P. Krishnasamy, Fiber and matrix-level damage detection and assessments for natural fiber composites, *Journal of Materials Science*, vol. 59, no. 36, pp. 16836–16861, 2024.

[140] N. Alahmed, I. Ud Din, W. J. Cantwell, R. Umer and K. A. Khan, Multi-scale characterization of self-sensing fiber reinforced composites, *Sensors and Actuators A: Physical*, vol. 379, pp. 115857, 2024.

[141] M. Liu, W. Saman and F. Bruno, Development of a novel refrigeration system for refrigerated trucks incorporating phase change material, *Applied Energy*, vol. 92, pp. 336–342, 2012.

[142] O. Adekomaya, T. Jamiru, R. Sadiku and Z. Huan, Minimizing energy consumption in refrigerated vehicles through alternative external wall, *Renewable and Sustainable Energy Reviews*, vol. 67, pp. 89–93, 2017.

[143] J. A. Casado, F. Gutiérrez-Solana, I. Carrascal, S. Diego, J. A. Polanco and D. Hernández, Fatigue behavior enhancement of short fiber glass reinforced polyamide by adding phase change materials, *Composites Part B: Engineering*, vol. 93, pp. 115–122, 2016.

[144] M. J. Kreder, J. Alvarenga, P. Kim and J. Aizenberg, Design of anti-icing surfaces: smooth, textured or slippery?, *Nature Reviews Materials*, vol. 1, no. 15003, 1, 2016.

[145] K. Zhu, X. Li, J. Su, H. Li, Y. Zhao and X. Yuan, Improvement of anti-icing properties of low surface energy coatings by introducing phase-change microcapsules, *Polymer Engineering & Science*, vol. 58, no. 6, pp. 973–979, 2018.

[146] A. M. Goitandia, M. B. Miguel, A. M. Babiano and O. G. Miguel, "Use of phase change materials to delay icing or to cause de-icing in wind-driven power generators," 2018.

[147] F. Chu, Z. Hu, Y. Feng, N. C. Lai, X. Wu and R. Wang, Advanced Anti-Icing Strategies and Technologies by Macrostructured Photothermal Storage Superhydrophobic Surfaces, *Advanced Materials*, vol. 36, no. 31, pp. e2402897, Aug 2024.

[148] V. D. Cao, *et al.* Influence of microcapsule size and shell polarity on thermal and mechanical properties of thermoregulating geopolymer concrete for passive building applications, *Energy Conversion and Management*, vol. 164, pp. 198–209, 2018.

[149] M. Yadav, N. Pasarkar, A. Naikwadi and P. Mahanwar, A review on microencapsulation, thermal energy storage applications, thermal conductivity and modification of polymeric phase change material for thermal energy storage applications, *Polymer Bulletin*, vol. 80, no. 6, pp. 5897–5927, 2022.

[150] F. Chen and M. Wolcott, Polyethylene/paraffin binary composites for phase change material energy storage in building: A morphology, thermal properties, and paraffin leakage study, *Solar Energy Materials and Solar Cells*, vol. 137, pp. 79–85, 2015.

[151] F. Chen and M. P. Wolcott, Miscibility studies of paraffin/polyethylene blends as form-stable phase change materials, *European Polymer Journal*, vol. 52, pp. 44–52, 2014.

[152] K. Resch-Fauster, F. Hengstberger, C. Zauner and S. Holper, Overheating protection of solar thermal façades with latent heat storages based on paraffin-polymer compounds, *Energy and Buildings*, vol. 169, pp. 254–259, 2018.

[153] P. Sobolciak, M. Mrlík, M. A. Al-Maadeed and I. Krupa, Calorimetric and dynamic mechanical behavior of phase change materials based on paraffin wax supported by expanded graphite, *Thermochimica Acta*, vol. 617, pp. 111–119, 2015.

[154] P. Sobolciak, M. Karkri, M. A. Al-Maaded and I. Krupa, Thermal characterization of phase change materials based on linear low-density polyethylene, paraffin wax and expanded graphite, (in English), *Renewable Energy*, Article vol. 88, pp. 372–382, Apr 2016.

[155] W. Wu, W. Wu and S. Wang, Form-stable and thermally induced flexible composite phase change material for thermal energy storage and thermal management applications, *Applied Energy*, vol. 236, pp. 10–21, 2019.

[156] G. Fredi, A. Dorigato, L. Fambri and A. Pegoretti, Wax confinement with carbon nanotubes for phase changing epoxy blends, *Polymers*, vol. 9, no. 9, pp. 405/1–16, 2017.

[157] X. X. Zhang, X. C. Wang, X. M. Tao and K. L. Yick, Energy storage polymer/MicroPCMs blended chips and thermo-regulated fiber, *Journal of Materials Science*, vol. 40, pp. 3729–3734, 2005.

[158] I. Krupa, *et al*. Phase change materials based on high-density polyethylene filled with microencapsulated paraffin wax, *Energy Conversion and Management*, vol. 87, pp. 400–409, 2014.

[159] J.-F. Su, X.-Y. Wang, S.-B. Wang, Y.-H. Zhao, K.-Y. Zhu and X.-Y. Yuan, Interface stability behaviors of methanol-melamine-formaldehyde shell microPCMs/epoxy matrix composites, *Polymer Composites*, vol. 32, no. 5, pp. 810–820, 2011.

[160] J.-F. Su, Y.-H. Zhao, X.-Y. Wang, H. Dong and S. B. Wang, Effect of interface debonding on the thermal conductivity of microencapsulated-paraffin filled epoxy matrix composites, (in English), *Composites Part A-Applied Science and Manufacturing*, Article vol. 43, no. 3, pp. 325–332, Mar 2012.

[161] X.-Y. Wang, J.-F. Su, S.-B. Wang and Y.-H. Zhao, The effect of interface debonding behaviors on the mechanical properties of microPCMs/epoxy composites, *Polymer Composites*, vol. 32, no. 9, pp. 1439–1450, 2011.

[162] G. Peng, *et al*. Enhanced mechanical properties of epoxy composites embedded with MF/TiO2 hybrid shell microcapsules containing n-octadecane, *Journal of Industrial and Engineering Chemistry*, vol. 110, pp. 414–423, 2022.

[163] G. Fredi, *et al*. Bioinspired Polydopamine Coating as an Adhesion Enhancer Between Paraffin Microcapsules and an Epoxy Matrix, *ACS Omega*, vol. 5, pp. 19639–19653, 2020.

[164] G. Fredi, C. Zimmerer, C. Scheffler and A. Pegoretti, Polydopamine-Coated Paraffin Microcapsules as a Multifunctional Filler Enhancing Thermal and Mechanical Performance of a Flexible Epoxy Resin, *Journal of Composites Science*, vol. 4, no. 4, pp. 174, 2020.

[165] C. Zimmerer, *et al*. Dopamine as a bioinspired adhesion promoter for the metallization of multi-responsive phase change microcapsules, *Journal of Materials Science*, vol. 57, pp 16755–16775, 2022.

[166] R. Wirtz, A. Fuchs, V. Narla, Y. Shen, T. Zhao and Y. Jiang, A Multi-Functional Graphite/Epoxy-Based Thermal Energy Storage Composite for Temperature Control of Sensors and Electronics, Reno Reno, Nevada, 89557 USA: University of Nevada, 2003, pp. 1–9.

[167] O. Mesalhy, K. Lafdi and A. Elgafy, Carbon foam matrices saturated with PCM for thermal protection purposes, *Carbon*, vol. 44, no. 10, pp. 2080–2088, 2006.

[168] Y. Zhong, Q. Guo, S. Li, J. Shi and L. Liu, Heat transfer enhancement of paraffin wax using graphite foam for thermal energy storage, *Solar Energy Materials and Solar Cells*, vol. 94, no. 6, pp. 1011–1014, 2010.

[169] P. Jana, V. Fierro, A. Pizzi and A. Celzard, Thermal conductivity improvement of composite carbon foams based on tannin-based disordered carbon matrix and graphite fillers, *Materials & Design*, vol. 83, pp. 635–643, 2015.

[170] S. Yoo, E. Kandare, R. Shanks, M. A. Al-Maadeed and A. Afaghi Khatibi, Thermophysical properties of multifunctional glass fibre reinforced polymer composites incorporating phase change materials, *Thermochimica Acta*, vol. 642, pp. 25–31, 2016.

[171] S. Yoo, E. Kandare, G. Mahendrarajah, M. A. Al-Maadeed and A. A. Khatibi, Mechanical and thermal characterisation of multifunctional composites incorporating phase change materials, *Journal of Composite Materials*, vol. 51, no. 18, pp. 2631–2642, 2016.

[172] S. Yoo, E. Kandare, R. Shanks and A. A. Khatibi, Viscoelastic characterization of multifunctional composites incorporated with microencapsulated phase change materials, in *International Conference of Materials Processing and Characterization (ICPMC)*, 2017.

[173] G. Fredi, A. Dorigato, L. Fambri, S. H. Unterberger and A. Pegoretti, Effect of phase change microcapsules on the thermo-mechanical, fracture and heat storage properties of unidirectional carbon/epoxy laminates, *Polymer Testing*, vol. 91, pp. 106747/1–106747/16, 2020.

[174] G. Fredi, A. Dorigato, L. Fambri and A. Pegoretti, Detailed experimental and theoretical investigation of the thermo-mechanical properties of epoxy composites containing paraffin microcapsules for thermal management, *Polymer Engineering and Science*, vol. 60, pp. 1202–1220, 2020.

[175] G. Fredi, A. Dorigato, S. Unterberger, N. Artuso and A. Pegoretti, Discontinuous carbon fiber/polyamide composites with microencapsulated paraffin for thermal energy storage, *Journal of Applied Polymer Science*, vol. 136, no. 16, pp. 47408/1–47408/14, 2019.

[176] G. Fredi, A. Dorigato and A. Pegoretti, Multifunctional glass fiber/polyamide composites with thermal energy storage/release capability, *Express Polymer Letters*, vol. 12, no. 4, pp. 349–364, 2018.

[177] A. Dorigato, G. Fredi, M. Negri and A. Pegoretti, Thermo-mechanical behaviour of novel wood laminae-thermoplastic starch biodegradable composites with thermal energy storage/release capability, *Frontiers in Materials*, vol. 6, pp. 1–12, 2019.

[178] G. Fredi, E. Boso, A. Sorze and A. Pegoretti, Multifunctional sandwich composites with optimized phase change material content for simultaneous structural and thermal performance, *Composites Part A: Applied Science and Manufacturing*, vol. 186, 108382, 2024.

[179] G. Fredi, A. Dorigato, L. Fambri and A. Pegoretti, Evaluating the multifunctional performance of structural composites for thermal energy storage, *Polymers*, vol. 13, no. 18, pp. 3108, 2021.

[180] G. Fredi, A. Dorigato and A. Pegoretti, Novel reactive thermoplastic resin as a matrix for laminates containing phase change microcapsules, *Polymer Composites*, vol. 40, no. 9, pp. 3711–3724, 2019.

[181] G. Fredi, A. Dorigato and A. Pegoretti, Dynamic-mechanical response of carbon fiber laminates with a reactive thermoplastic resin containing phase change microcapsules, *Mechanics of Time-Dependent Materials*, vol. 24, no. 3, pp. 395–418, 2020.

[182] G. Fredi, A. Dorigato, L. Fambri and A. Pegoretti, Multifunctional epoxy/carbon fiber laminates for thermal energy storage and release, *Composites Science and Technology*, vol. 158, pp. 101–111, 2018.

[183] A. Dorigato, G. Fredi and A. Pegoretti, Application of the thermal energy storage concept to novel epoxy/short carbon fiber composites, *Journal of Applied Polymer Science*, vol. 136, no. 21, pp. 47434/1–47434/9, 2019.

[184] D. J. O'Brien, D. M. Baechle and E. D. Wetzel, Design and performance of multifunctional structural composite capacitors, *Journal of Composite Materials*, vol. 45, no. 26, pp. 2797–2809, 2011.

Index

https://doi.org/10.1515/9783111111865-008